The Early Development of Radio in Canada
1901-1930

The Early Development of Radio in Canada
1901-1930

An Illustrated History of Canada's Radio Pioneers, Broadcast Receiver Manufacturers and their Products

Edited by Robert P. Murray

Chandler, Arizona

~ for Eileen ~

Sonoran Publishing, LLC, Chandler, Arizona 85226
©2005 by Robert P. Murray. All rights reserved.
First printing June 2005

Printed in the United States of America

Library of Congress Cataloging-in-Publication Data

    The early development of Radio in Canada, 1901-1930: an illustrated history of Canada's radio pioneers, broadcast receiver manufacturers, and their products / edited by Robert P. Murray.
        p. cm.
Includes bibliographical references.
        ISBN 1-886606-20-X (pbk.)
  1. Radio—Canada—History. 2. Radio—Canada—Equipment and supplies—History. 3. Radio broadcasting—Canada—History. 1. Murray, Robert P., 1944-

TK6548.C2E27 2005
384.54'0971—dc22

On the Cover:
This circa 1901 photo shows Guglielmo Marconi reading a paper tape of received messages, and his assistant George Kemp poised to operate the transmitting key. Degna Marconi in her book *My Father, Marconi* (New York: McGraw Hill, 1962) describes it as a scene from the first film about wireless made by the Biograph Company shortly after the 1901 demonstration. Peter R. Jensen in his book *In Marconi's Footsteps: Early Radio* (Kenthurst NSW Australia: Kangaroo Press, 1994) describes it as Marconi with Kemp and the wireless station at Dover, 1899.

Contents

	Foreward	vi
	Acknowledgements	vii
	Introduction	viii
1	**Reginald Fessenden and the National Electric Signaling Company (NESCO)**	1
	1.1 Reginald Fessenden's Liquid Barretter	
2	**The Canadian Marconi Company**	7
	2.1 The First Thirty Years of the Canadian Marconi Company	
	2.2 The Development of Radio in Canada to 1928	
3	**The Telephone Industry**	47
	3.1 Manufacture of Broadcast Receivers by the Northern Electric Company in the 1920's	
	3.2 Radios of the Canadian Independent Telephone Company	
4	**Rogers Radio Ltd.**	69
	4.1 "Just Plug In—Then Tune In" — The First Commercial Light-Socket Operated Radio Receivers With AC tubes from Rogers Radio Ltd., Toronto, Canada	
5	**Patents in Radio**	97
	5.1 Patents and the radio industry in Canada	
6	**W.W. Grant Radio**	103
	6.1 "The Voice of the Prairie" — A Brief History of W.W. Grant (1892-1968)	
7	**Patent Avoidance — Kit Radios**	109
	7.1 Eaton Single Circuit Set, 1923	
8	**The Mercury Super Ten**	113
	8.1 The Mercury Super Ten	
	8.2 Before and after the Mercury Super Ten	
9	**The Component Parts of the Radio Corporation of America**	121
	9.1 Broadcast Receiver Manufacture by General Electric and Westinghouse in the First Decade of RCA	
	9.2 Major C.L. Richardson, Canadian Radio Engineer	
10	**Starting a Radio Company in the 1940's**	143
	10.1 Hank Thorkelsson and the Thorcraft Radio Co.	
	Epilogue	149
	Index	151

Foreword

The early history of radio in Canada is a story rarely told. Too often it gets abbreviated to a short account of Guglielmo Marconi's electrifying, if somewhat equivocal, transoceanic signal reception in St. John's, Newfoundland in December 1901. As other historians have pointed out, the barely audible results of this experiment merely marked the beginning of a long and arduous process of technical refinement and entrepreneurial improvisation that would take three decades to reach fruition. The history of these decades is what Dr. Robert Murray offers us in *The Early Development of Radio in Canada, 1901-1930*.

By the time of the First World War, marine radio was on a firm footing on Canada's coasts and inland waters. This was based on Marconi's spark gap technology, which proved adequate for the few high powered coastal and shipboard radiotelegraph stations that were using it. Already though, research by the Canadian scientist Reginald Fessenden had demonstrated the limitations of this technology and had pointed to the future: transmission and reception of voice and code by continuous waves. The next few years were consumed in war, but also in hammering out the practical details of Fessenden's system, particularly the perfection of vacuum tube oscillators and amplifiers and the development of workable transmission and reception circuits.

As the 1920s dawned, the essential pieces for an enormous expansion in the scope and impact of radio were in place. As the phenomenon of broadcast listening emerged out of the esoteric hobby of amateur radio, several companies undertook the manufacture of receiving equipment for this growing market. Some, like the Canadian Independent Telephone Co., W.W. Grant and H.M. Kipp, have long since disappeared. Others, like Canadian Marconi, Northern Electric, Rogers Radio and Canadian General Electric still exist, though frequently in altered corporate form. The receivers of the 1920s assumed a multitude of forms — regenerative, TRF, neutrodyne, superheterodyne — as builders sought to balance manufacturing cost with simplicity of operation and the avoidance of competitors' patents. By 1930 the situation had stabilized. The superheterodyne reigned supreme and a handful of large companies were pooling their patents to dominate the market for receiver sets.

This book is a collection of stories, and I can think of no one better suited to bringing them together than Bob Murray. I first met Bob in the late 1990s. In 2000 he showed me his superb collection of Canadian broadcast receivers. In a generous gift to posterity, Bob subsequently donated a large portion of this collection to the Canada Science and Technology Museum. At the time, Bob told me his interests were shifting from collecting to historical research. This book is a result of that new passion.

The Early Development of Radio in Canada, 1901-1930, assembles in one volume 14 essays on the early years of Canada's radio industry. Until recently, the history of science and technology in Canada has not received a lot of attention from professional historians. In the radio field, I can think of only one book with a national scope, Sharon Babaian's *Radio Communication in Canada* (1992). We are therefore indebted to passionate collectors and amateur historians for much of what we now know about Canadian radio history. In 2002 Lloyd Swackhammer published his encyclopedic study of broadcast receivers, *Radios of Canada*. Now Bob Murray and his colleagues have given us *The Early Development of Radio in Canada, 1901-1930*, offering a wealth of knowledge about the individuals and organizations who pioneered wireless technology in this country.

Though the essays that follow have mostly been previously published in various periodicals, they achieve a greater value and interpretive power when now brought together in a single volume. I have no doubt anyone interested in radio history will be engrossed by the texts and fascinated by the many photos reproduced here. They will also be impressed by the dedication and scholarship that went into producing these histories. Like all good books, *The Early Development of Radio in Canada, 1901-1930*, will become the starting point for future study. No longer lost in back issues of collector magazines and newsletters, I hope that as a result the essays will be "broadcast" to a new and larger audience.

Bryan Dewalt
Curator of Communications
Canada Science and Technology Museum, Ottawa
2 December 2004

Acknowledgements

Much of the content of this book is reprinted by permission from scattered sources. A person interested in this history would not normally have access to all of these articles, and some of them have been updated. The sources listed are predominantly associations of amateur radio historians and collectors of early radio artefacts. Without their generous policies about reproduction of previously published material, this book would have not been possible.

Article 1.1 is reprinted from the *AWA Journal* by permission of the Antique Wireless Association, Inc., Box E, Breesport, NY, 14896, USA. It is, 1.1. Murray, R. Reginald Fessenden's liquid barretter. 2005; vol. 46(2): pp. 37-39, and vol. 46(3).

Articles 6.1 and 8.2 are reprinted from *The Old Timer's Bulletin* by permission of the Antique Wireless Association, Inc., Box E, Breesport, NY 14816, USA. They are, **6.1**. Murray, R. "The voice of the prairie": A brief history of W.W. Grant (1892–1968). 1992; vol. 33(3): pp. 16-19. **8.2**. Murray, R. Before and after the Mercury Super Ten. 1994; vol. 35(4): pp. 10-13.

Articles 2.1, 3.1, 4.1 and 9.1 are reprinted from the *AWA Review* by permission of the Antique Wireless Association, Inc., Box E, Breesport, NY 14816, USA. They are, **2.1**. Hart, R., and Murray, R. The first thirty years of the Canadian Marconi Company. 2001; vol. 14: pp. 92-147. **3.1**. Murray, R. Manufacture of broadcast receivers by the Northern Electric Company in the 1920's. 2000; vol. 13: pp. 7-37. **4.1**. Chaplin, M. "Just plug in—then tune in" – The first commercial light-socket operated radio receivers with AC tubes from Rogers Radio Ltd., Toronto, Canada. 2002; vol. 15: pp. 147-177. **9.1**. Murray, R. Broadcast receiver manufacture by General Electric and Westinghouse in the first decade of RCA. 2004; vol. 17: pp. 107-142. Subsequent reproduction of any of this material will require additional permission from the Antique Wireless Association, Inc.

Articles 2.2 and 5.1 are reprinted from the *Ottawa Vintage Radio Club Newsletter*, 26 Ossington Avenue, Ottawa ON K1S 3B4, by permission. They are, **2.2**. A.F. Fraser, The development of radio in Canada to 1928. 1997, Winter, pp. 3-5, 1998; Spring: pp. 3-4, and 1998; Fall: pp. 3-6. **5.1**. Symonds, G. Patents and the radio industry in Canada: Developments to the end of the First World War. 2000; Fall: pp. 13-20.

Articles 3.2 and 7.1 are reprinted from *Antique Radio Classified*, P.O. Box 2, Carlisle, MA 01741, USA. These are copyright John V. Terrey and are reprinted with permission. They are, **3.2**. Murray, R. and Frederickson, H.K. Radios of the Canadian Independent Telephone Co., Limited. 2000; vol. 17(3): pp. 4-7. **7.1**. Murray, R. Eaton single circuit set, 1923. 1994; vol. 11(8): pp. 4-5.

Article 9.2 is reprinted from *The Cat's Whisker*, Bulletin of the Canadian Vintage Wireless Association, 38 Grenview Blvd., N., Toronto, ON, M8X 2K2. The article is **9.2**. Challoner, D.A. Major C.L. Richardson, Canadian radio engineer. 1978; vol. 8(3): pp. 6-17.

Article 10.1 is reprinted from *Radio Waves*, the Official Publication of the Canadian Vintage Radio Society, 4895 Mahood Drive, Richmond, BC V7E 5C3. The article is **10.1**. Murray, R. Hank Thorkelsson and the Thorcraft Radio Co. 1994; vol. 2(6): pp. 8-11. A supplement was printed in 1999; vol. 7(1) pp. 8-10, and is included in article 10.1.

Articles 2.2 and 9.2 which appeared in *The Cat's Whisker* were supplied in machine readable form by Gordon Symonds to whom we are sincerely grateful.

Finally, I wish to acknowledge the encouragement and assistance of Bryan Dewalt, Curator of Communications, Canada Science and Technology Museum, Ottawa. Also I appreciate the detailed and painstaking help of my publisher George A. Fathauer of Sonoran Publishing, LLC, and his confidence in the project expressed in his acceptance of this book for publication.

Introduction

The purpose of this book is to document the stories of the origins of much of Canadian radio broadcast receiver manufacturing. It aims to give the flavor of the wireless communications environment as well as the actions of the companies that responded to it. It covers the period from the beginnings of practical wireless communication in 1901 until the end of the first decade of broadcasting around 1930. The emphasis is on the invention and production of early apparatus and not on broadcasting. The predominant inventions and developments in radio were created elsewhere and not in Canada, with some notable exceptions. The story of early developments in radio as they impacted Canada, and of the people involved, has not previously been told.

Because of the origins of this material in publications of amateur historian/collectors of early radio, its focus is a blend of reporting early people and events, and describing early apparatus. This book does not attempt to be complete in its coverage of radio receivers made in Canada. This aim was addressed ably by the author of an earlier book, *Radios of Canada* (published privately in 2002 by Lloyd Swackhammer, R.R. #2, Alma ON N0B 1A0).

The radio industry in the early days was characterized by both the careful actions of the major corporations and the flamboyant actions of individual entrepreneurs. The larger players essentially guided where the industry was likely to go, and the minor players provided much of the sense of excitement. As we shall see in the chapters that follow, the participating major corporations were most likely to survive and diversify into other areas of business. The minor companies were more likely to disappear.

If there is a case where a few entrepreneurs at the beginnings of radio evolved into a major corporation long afterward, that case would be Rogers Radio Ltd. The company began in the "Radio Boom", the beginning of the popularity of broadcasting. Looking backward from today in 2005, it is hard to imagine those tentative first steps; their story is included here.

I hope this book will serve collector's purposes, and those of readers with an interest in technical history. It has been a pleasure to write it.

~ 1 ~
Reginald Fessenden and the National Electric Signaling Company (NESCO)

Although much of his important work precedes broadcasting, in some ways he seems a suitable place to start in reporting Canadian developments. Fessenden was born in Canada, although he later became an American citizen. He was educated first in Canada, even though the informal nature of much of his education makes it difficult to trace to a location. He attended university, and for eight years was head of electrical engineering at Purdue University and then the Western University of Pennsylvania, but never received a university degree himself.

It seems at least partly justifiable to claim some of Fessenden's achievements for Canada, and for our convenience in this book we shall do so. As this chapter describes, Fessenden recognized the need to shift the technology of wireless communication from spark transmission to continuous wave transmission. With the earliest forms of such continuous wave transmission apparatus he sent the first voice and music along the eastern seaboard of the U.S. and the Caribbean. Wireless operators on ships of the United Fruit Company, equipped with his receiving apparatus, were amazed at what they heard. Until that time, all wireless communications had been in Morse code.

Reginald Fessenden's liquid barretter
Robert Murray

Reginald Fessenden was born in 1866 in Knowlton, Canada East (now Quebec). He was precocious as a child, and at age 16 was hired to teach mathematics at Bishops College School, Lennoxville, Quebec. Four years later he was principal of Whitney Institute, a small school in Bermuda. From 1887-1890 he worked at Thomas Edison's Laboratory, and from 1890-1892 he worked for Westinghouse in Pittsburg. In 1892 he was selected as chair of electrical engineering at Purdue University although he held no university degree, and a year later he was recruited to the same position at the Western University of Pennsylvania (later the University of Pittsburgh) through the influence of George Westinghouse.

In 1900 Fessenden (Figure 1) joined the U.S. Weather Bureau to help them transmit weather forecasts, but soon fell out with his supervisor and left in 1902. With two Pittsburgh millionaires, T.H. Given and Hay Walker, he formed the National Electric Signaling Company (NESCO), a firm which survived for the next ten years.[1]

Fessenden's inventions were guided by a strong grasp of both theory and mathematics. He was the first inventor who saw that signaling needed to change from spark technology to continuous wave technology.[2] He wrote the specification from which Ernst Alexanderson at G.E. developed the famous alternator, which for the first time was conceived as a transmitter itself, not just as a source of power.

Fig. 1. Reginald Fessenden

The receiver in Fessenden's system was, he claimed, a thermo-electric one he called the *liquid barretter* and was patented in 1903. Signals were detected at a heated junction between platinum wire and nitric acid. The heating aided the rectification at the junction, and was augmented by a bias battery. However, in a fairly recent study of electrodes in an electrolyte solution, Geddes et al. make no mention of the temperature of the electrolyte as a contributor to its rectification properties.[3] In a report to the Smithsonian Institution in 1908, Fessenden conceded that electrolytic detectors invented by others also worked, but his worked better. He said that perhaps he was just more familiar with operating his own![4] The courts awarded the *liquid barretter* patents to Fessenden. In the decade between 1903 and 1913, the *liquid barretter* was the predominant continuous wave detector in commercial use. Apparently none remain, the one at the Ford Museum in Dearborn MI being a replica (Figure 2).

Fessenden's pioneering voice transmission in December 1906 was made with a wireless system that included this device.

Method

Because no detailed plans remain describing the *liquid barretter*, I decided to make a working model rather than attempt a look-alike model. I relied most heavily on the information in Fessenden's U.S. patent no. 727,331 dated May 5, 1903. It specified a fine pointed platinum wire of 0.0004 inch diameter just touching the surface of nitric acid in a glass vessel. A second platinum electrode was formed into a small coil (see FIG. 3 in Figure 3).

The fine platinum wire was Wolloston wire, 0.002 inch platinum plated first with silver and then further drawn to the smaller diameter. The silver was etched off by acid. I chose instead to work with 0.001 inch platinum wire now available (Surepure Chemetals Inc., Florham Park, NJ). I doubted whether Fessenden had at his disposal the equipment necessary to measure the final diameter of his platinum wire. I made a stand capable of holding a glass cup of acid, of clamping both electrodes in place, and of raising and lowering the anode. The base was a block of 6 x 7 x 1.5 inch solid Honduras mahogany finished with shellac. The bias voltage in my model was controlled externally, whereas in the *liquid barretter* I suspect it was controlled internally. The anode was of 0.001 inch platinum wire as stated, and the cathode was a ½ inch length of 1/8 inch diameter carbon rod removed from an AAA-size

Fig. 2. Liquid barretter in the Ford Museum in Dearborn, MI. (Larry Babcock photo.)

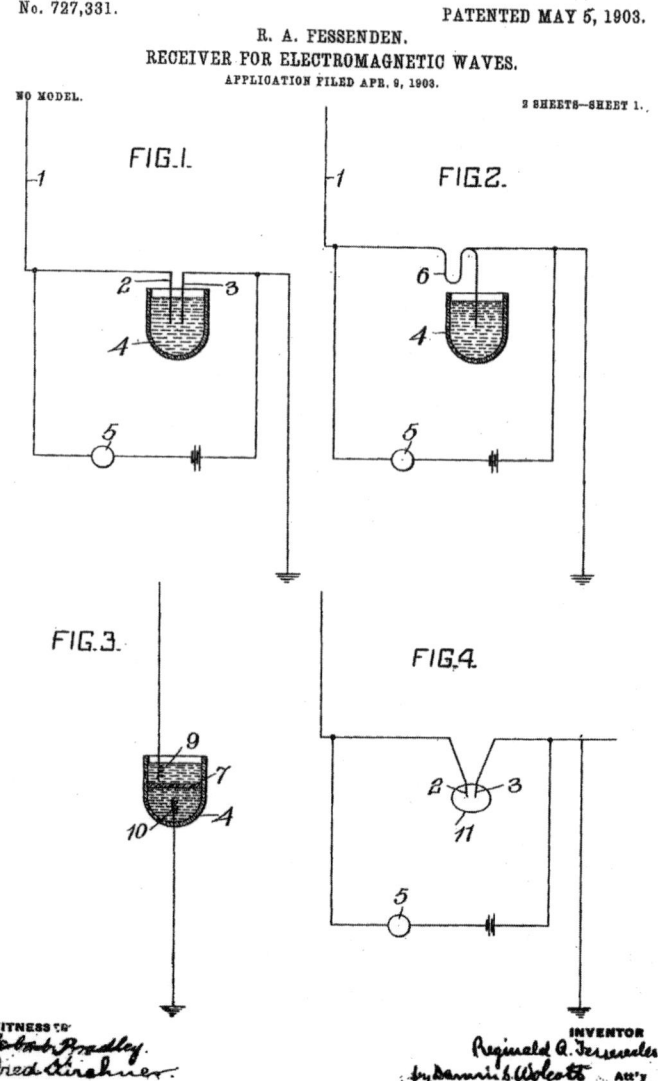

Fig. 3. Drawings from U.S. patent 727,331 granted May 5, 1903.

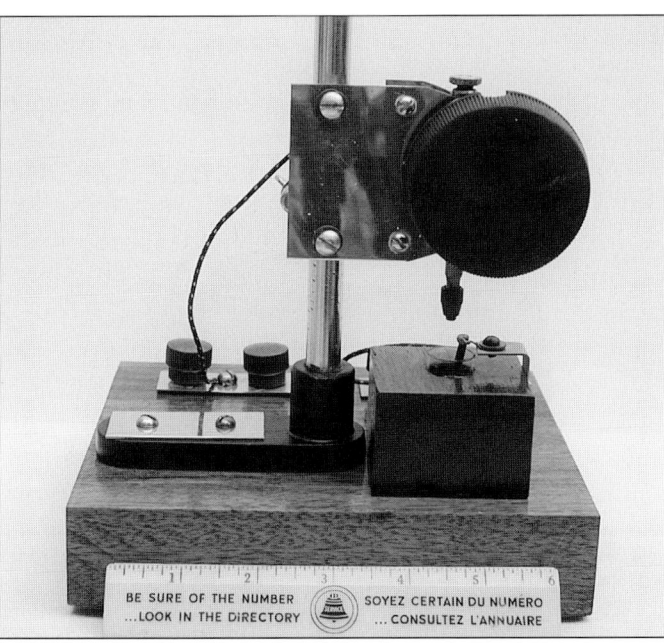

Fig. 4. Model of Fessenden's liquid barretter which did not work.

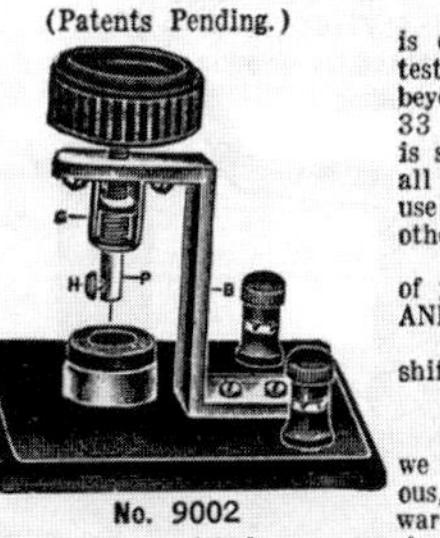

The Bare Point "Electro"-Lytic Detector

(Patents Pending.)

No. 9002

There is no doubt but that the Bare Point Electrolytic is one of the most sensitive responders invented. Comparative tests by the foremost men in the wireless field have proven beyond doubt that the Electrolytic Detector is fully 25% to 33 1/3% more sensitive than any mineral detector. That this is so is best proven by calling attention to the fact that nearly all up-to-date commercial and government wireless stations now use the Electrolytic type, as its range is far greater than any other detector in existence, except the Radioson.

Our No. 9002 Detector has been on the market for a score of years and is, despite its low price, unquestionably the BEST AND MOST EFFICIENT TYPE on the market.

Furthermore this instrument is not an amateurish make-shift, but has been designed for hard commercial work.

DESCRIPTION.

To hold the acid we use a graphite-carbon cup, which we absolutely guarantee will never leak; this cup is non-porous, being "formed" under a pressure of 300 tons, and afterwards treated in the furnace to close up any microscopic holes. The acid will positively never soak into the cup. Ordinary carbon or graphite—as anybody can readily convince himself by a trial—will absorb the acid in less than ten minutes. Base is 3½x3½x¼ in. size made of molded hard rubber. The standard is nickel plated and polished as are all the metal parts of the detector.

We use a very small graphite cup, to hold but little acid. Several reasons prompted us for doing so. First, a small body of liquid will adhere to the container better than a large one. Shocks and unsteady position (on ships) will therefore effect our instrument but little. Secondly, the smaller the surface of the liquid, the slower the evaporation.

There could be scarcely a better method of establishing contact with the graphite-carbon cup than the one we have adopted. The cup is forced into a brass cap under hydraulic pressure, so that three-quarters of the cup is actually in contact with the brass cap.

The most important part of an Electrolytic Detector, however, is found in the mechanism used to raise and lower the fine Wollaston wire.

The ideal way, of course, is to move the wire up and down strictly vertical, so that one can dip it more or less in the electrolyte (acid). This method has been adopted in our new instrument. It is unquestionably an ingenious movement.

Chuck P, holding the fine wire, has at the upper end a square guiding rest, under which a spring is located which tends to press P upward. This is counteracted by the adjusting thumb-screw, which, by pressing on the flat part on P, raises or lowers same at will. It is impossible for the pin P to turn, as the square guiding rest can only move up or down in the guiding angle G. The large hard-rubber thumb-screw prevents current leakage through body; it allows extremely fine adjustments, and gives the instrument a distinctive appearance. The screw, having a very fine pitch, makes it possible to raise or lower the fine wire less than 1-1000 inch at a time. This allows very fine adjustments.

The nickel-plated bracket B is of generous proportions and unusually heavy in order to stand rough and continuous handling.

The Wollaston wire W used is the very thinnest platinum wire obtainable. Its diameter is 0.0001 inch. Wollaston wire comes heavily silver-plated. The acid soon dissolves the silver, leaving only the minute platinum wire of 1-10,000 inch diameter.

The best part, however, is that it is unnecessary to solder this fine wire, which method, not alone inconvenient to many, is too troublesome and takes too much time. Pin P at the lower part has a fine hole ½ inch long, into which the wire is inserted, then the small thumb-screw H is tightened and the wire is secure.

We also furnish (with the instrument only) a novel wrinkle how to handle and insert the fine wire in the hole, without the screw H damaging the fine wire in the least, even if screwed up quite tight. Using this method it is possible to use the last ⅛ inch piece of wire, which heretofore was impossible. One inch of wire therefore will last a long while.

The base (size 3½x3½ in.) is ¼ in. thick, being of molded hard rubber composition. Cannot shrink or warp. It gives the Detector a business-like appearance, and is, of course, acid proof. Two binding posts are provided.

We cannot send the acid through the mails or by express. We furnish rubber-stoppered bottle, which bears formula how to prepare the electrolyte. It can be obtained at ANY drug store. **Formula is only given with order.**

One of our "Electro" Potentiometers should always be used when a battery is in circuit, else the strong current will soon dissolve the minute platinum wire of the Detector. The Potentiometer further eliminates all humming noises sometimes experienced in the telephone receiver.

Gentlemen:— **Oshawa, Ont., Can.**

I received your Radioson Detector and am very pleased with it.

GEO. B. NORRIS.

16

Fig. 5. Page from 1914 E.I. Co. catalog showing electrolytic detector.

lead-acid battery (see Figure 4). The anode was held in place in a jeweler's pin vise.

Fessenden specified nitric acid, although others have mentioned sulphuric acid. Fessenden never indicates the acid concentration although others have suggested 20%. I tested my detector with a SkyWaves crystal set designed by Al Klase and described at the 1999 AWA conference, and on his web site.[5] Since I had built this model for presentation at the AWA annual conference in August, 2004, I ended up rushing its completion. This detector failed to work.

I tried a simpler detector stand as the conference drew nearer. I fashioned one after that shown in the 1914 E.I. Co. catalog, see Figure 5. This had a similar arrangement of electrodes in a glass cup, but in this case the cathode was a coil of about 3 inches of the same platinum wire (Figure 6). The anode was about ¼ inch of platinum wire soldered into the advancing screw. This detector worked, but very faintly. I found that it worked best when the wire was just about drawn up out of the acid, and possibly pulling on the surface tension. I do not know why the device worked in this way. I mention this only as a speculation. I can hardly see 0.001 inch platinum wire without a magnifying glass. I can certainly not see when it dips into the acid, but can hear the result in the headphones. When I hold the wire between my fingers I can not feel it.

In comparison to the 1N34A diode in the SkyWaves crystal set, I measured the output of the *liquid barretter* (now just an electrolytic detector) at -12 dB using a 830 kHz AM carrier and 1 kHz modulation at 50%.

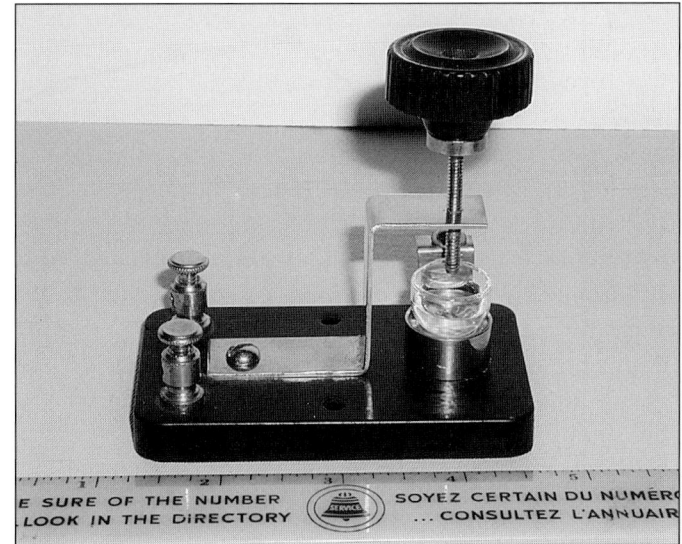

Fig. 6. Model electrolytic detector similar to the one in the 1914 E.I. Co. catalog. This model detected a signal faintly.

My experimentation with a DC bias yielded no positive effect. Artifacts and poster in hand, I went off to the AWA conference.

The conference aftermath

A clear benefit of presenting such a half-baked project was that I collected quite a number of useful suggestions from those present. I returned home and redesigned the simpler of my two detectors. I sacrificed a number 6 dry cell to obtain the 7/8 inch diameter carbon rod contained in it. Because I was reluctant to try turning the rod to a smaller diameter and because I wanted to leave a lot of carbon between myself and the acid, I left the diameter at 7/8 inches and cut off a ½ inch slice with a band saw. Then I carefully drilled a hole about ¼ inch deep with a sharp bit. Both the carbon disc and the drill were held in a Unimat lathe. I then turned a brass cup in which the carbon fit loosely and was held by a set screw. The cup was screwed to my detector stand (Figure 7). The feed screw with a short (about 3/8 inch) piece of platinum wire soldered to it was now held over the hole in the carbon disc. I added a threaded bushing below the tapped hole in the bracket to minimize the play in the adjustment.

I then could compare the performance of the electrolytic detector and the 1N34A across frequencies

Fig. 7. Model closer to the E.I. Co. detector. This one worked better.

(or, in terms used between 1900 and 1910, wavelengths). Using an RF signal generator (U.S. Navy model AN/URM-25H) I supplied 500 kHz (600 metres) to 1,000 kHz (300 metres) to the crystal set in increments of 100 kHz modulated 50% at 1 kHz (see Figure 8). These wavelengths were in common ship-to-shore use around 1905. Output was measured with a Hewlett-Packard model 403B dB meter. Each detector was tested several times and the best performance is shown in the Figure. Tests were also conducted at

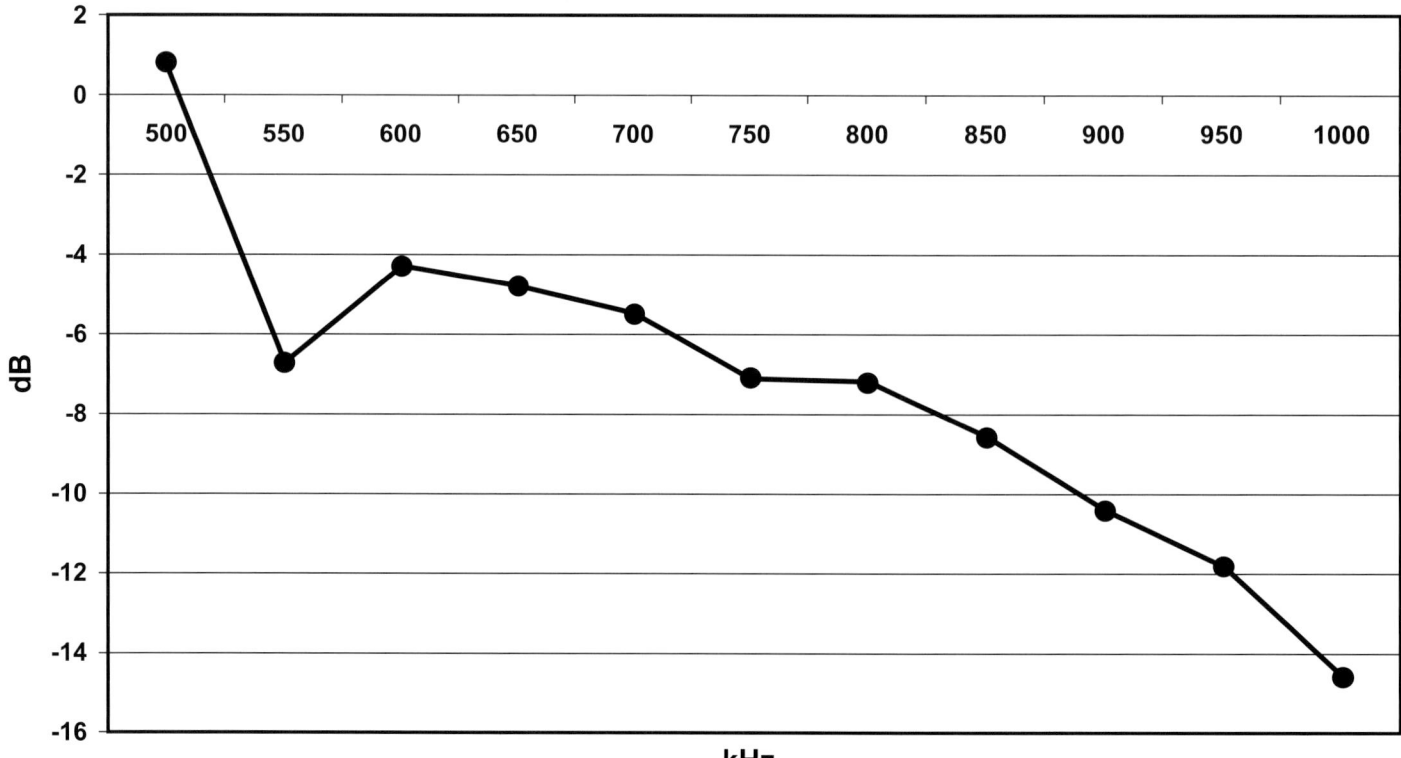

Fig. 8. Test of output with electrolytic detector minus output with 1N34A in the Skywaves receiver with signal modulated at 50% with 1 kHz.

modulation frequencies from 400 Hz to 1,200 Hz with each detector, and there was no effect of varying the audio frequency over this range.

The pattern of output values for the 1N34A across frequencies was reasonably flat, except for a drop at 550 kHz seen also with the electrolytic detector. This was likely a flaw in the tuning characteristics of the set. Apart from this, the electrolytic detector and the 1N34A were associated with about equal performance at 500 kHz, beyond which the output with the electrolytic detector fell to about 7 dB below the 1N34A at 750 kHz and to about 14 dB below at 1,000 kHz.

I was then able to compare the internal 1N34A diode with external galena, perikon (2), carborundum and electrolytic detectors. Using the internal 1N34A diode as a reference, the output obtained with other detectors at 750 kHz modulated 50% at 1 kHz is shown in Table 1.

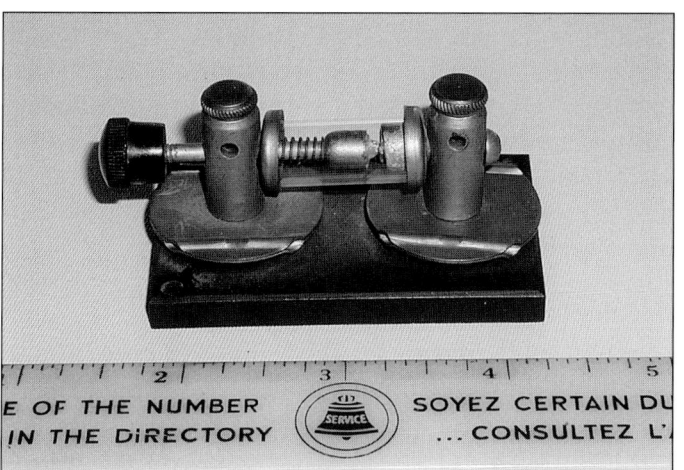

Fig. 9. Perikon detector of unknown manufacture

Fig. 10. Carborundum detector of unknown manufacture, with replica lever

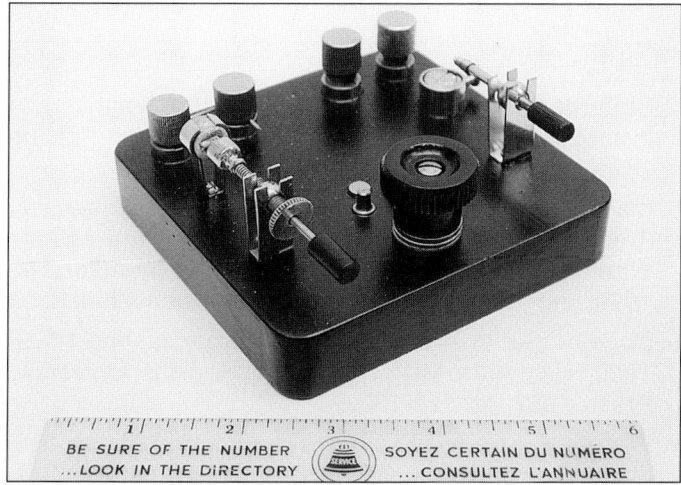

Fig. 11. Westinghouse type DB replica

Table 1. Output values of a SkyWaves crystal set measured in dB for five external detectors, with reference to the 1N34A diode in the set.

1N34A (reference)	0 dB
perikon #1 (unknown manufacture) Figure 9	-1.9 dB
carborundum (unknown manufacture) no bias, Figure 10	-5.2 dB
galena Westinghouse type DB replica, Figure 11	-5.4 dB
electrolytic no bias, Figure 7	-5.6 dB
perikon #2 Westinghouse type DB replica, Figure 11	-10.8 dB

Note that the output with the electrolytic detector at 750 kHz is shown as -7.1 dB in Figure 8 and as -5.6 dB in Table 1. Differences of this magnitude were often seen when separate series of measurements were compared, and are due to the multifactorial nature of the experimental setup, or else my lack of skill in operating it.

Both carborundum and electrolytic detectors are sometimes used with a small bias voltage across them. In this arrangement 3 or 4 volts DC is supplied through a 500 ohm potentiometer, and the adjustment is said to be critical. I was unable to demonstrate any improvement with the bias on either detector. With my eye on the dB meter, I found that there was a transient increase due likely to noise from the insertion. The dB subsequently returned to the earlier value. It occurred to me that without a sensitive output meter, it might be difficult to distinguish these events from an improvement in output.

Conclusions

After some trial and error, a satisfactory electrolytic detector was demonstrated although the addition of a DC bias did not improve its performance. Fessenden in 1903 argued that the action of his detector was thermoelectric rather than electrolytic, but he would not have had instruments necessary to indicate a heating effect, and more recent authors do not mention heating as part of the rectification phenomenon in liquids. Nonetheless Fessenden's *liquid barretter* worked, and he was awarded a number of patents for it. I suspect it was simply a form of electrolytic detector of which several were invented around the same time. Fessenden's *liquid barretter* cost $2,500 and the U.S. Navy refused to buy them from him.[6] Under these circumstances I'd have recommended the $1.25 E.I. Co. version. Aitken[2] observed that the secretary of the navy had stated that Fessenden was asking too high a price and therefore the department "feels that it is relieved of any moral obligation that might otherwise exist." Clearly it was the Navy's morality that had a price tag, not Fessenden's (p. 57).

In early books the electrolytic detector is described as superior to any mineral detector. The electrolytic detector was discussed at the time of the carborundum detector and the magnetic detector, the other materials having come later. In my tests, the carborundum, galena and electrolytic detectors are essentially indistinguishable in performance. Perikon detectors worked both better and worse than these. My rankings are difficult to compare to those conducted in 1915 and reported by Phillips in 1980.[7] Part of the reason for the difficulty is that in the era of spark transmission, a unique set of factors enters the comparison. A feature of the *liquid barretter* and electrolytic detector is that they span the time from the era of spark to the time of continuous wave transmission. Fessenden through NESCO had provided *liquid barretters* to United Fruit Company ships. Those ships in the Caribbean heard his first ever voice transmission in December of 1906. Subsequent detectors were developed with continuous waves in mind.

Acknowledgements

I thank Don Trueman for the loan of the test equipment and for his technical advice. Welcome suggestions were made by Larry Babcock, Brian Belanger, Bill Holley, Norm Hertz, Bart Lee, Robert MacIntyre, Peter Yanczer and others at the 2004 AWA conference.

REFERENCES

1. Belrose JS. Fessenden and the early history of radio science. Proceedings of the Radio Club of America, 1993; November: 6-23.
2. Aitken HGJ. The continuous wave: Technology and American radio, 1900-1932. Princeton, NJ: Princeton University Press, 1985.
3. Geddes LA, Foster KS, Reilly J, Voorhees WD, Bourland JD, Ragheb T, Fearnot NE. The rectification properties of an electrode-electrolyte interface operated at high sinusoidal current density. IEEE Transactions on Biomedical Engineering. 1987; BME-34:669-672.
4. Fessenden RA. Wireless telegraphy. Proceedings of the American Institute of Electrical Engineers, 1908;27. Reprinted in the Annual Report of the Smithsonian Institution, 1908. Washington DC: Government Printing Office, 1909, pp. 161-195.
5. Klase AR. The Skywave high-performance crystal set. Presentation at the Annual Conference of the Antique Wireless Association, Rochester NY, September 1-4, 1999. http://www.webex.net/~skywaves/home.htm.
6. Douglas SJ. Inventing American broadcasting 1899-1922. Baltimore MD: The Johns Hopkins University Press, 1987, pp. 128-129.
7. Phillips VJ. Early radio wave detectors. London: The Institution of Electrical Engineers, 1980, pp. 213-214.

~ 2 ~
The Canadian Marconi Company

In terms of its impact on the wireless communication environment in Canada, the company formed by Guglielmo Marconi in England was clearly the most influential in the early days. Marconi's domination of transatlantic communication first to St. John's, Newfoundland (not then part of Canada), and his decision to build his first working transatlantic station at Glace Bay, Nova Scotia, were key in the direction of later developments. Marconi's agreement with the Canadian government to be its sole provider of wireless facilities (then regarded as a ship-to-shore service) seems now to have been more a marketing achievement than an engineering one.

The view of the adoption of the Marconi system from the Canadian Government side is outlined in the following section. It is an excerpt from an address presented to the Engineering Institute of Canada by A.F. Fraser, Chief Engineer, Radio Branch, Department of Marine, 1930.

The first 30 years of the Canadian Marconi Company
Roger Hart and Robert Murray

This year, 2001, is the 100th anniversary of Guglielmo Marconi's report of reception of a wireless signal in Newfoundland from a transmitter in Poldhu, England. This event had among its many consequences the establishment of the Marconi Company in Canada. It seems particularly appropriate now to document the development of wireless telegraphy from the perspective of Marconi's Canadian company.

The early years of the Canadian company were documented by D.R.P. (Darby) Coats, among others. Coats was an employee who began his career by training in a wireless operator's school in his native England, and then being assigned as a ship's wireless operator under the direction of the Marconi office in Montreal. Later, in the 1920's he worked as public relations officer for the Canadian Marconi Company (CMC) in Montreal, and later was assigned (or perhaps took it upon himself) to write the history of the company. Roger Hart was Manager, Marketing Services with CMC from 1981 to 1996. He has been an active radio amateur (currently VA3OA) since 1964, and is a current AWA member. Some years ago, Roger

Darby Coats *was born in Gravesend, some 20 miles south of London. In 1910 Darby went to a school in Stockwell to learn to be a wireless operator. His contemporaries at the school included Harold "Judy" Bride, later assistant wireless operator on the Titanic, and Cyril "Corp" Evans, destined to be wireless operator on the Californian. Upon graduation, Darby was assigned to the Marconi Company office in Montreal, from which he was dispatched on ships sailing from that city (Figure 1). Coats moved into the CMC offices in 1919, so that he did some announcing in the early days of experimental station XWA. He also had some public relations duties with the firm. In early 1923 Coats had moved to CKY, a government owned broadcast station in Winnipeg, as an announcer and later as public relations director. He appears to have spent most of his career with CKY in Winnipeg, except for his service in the Air Force during WW II, and a brief assignment with the YMCA after the war. While at CKY in the late 1930's and 1940's, he published a listener's newsletter titled Manitoba Calling, which contained first hand accounts of the early days at CMC. It is not clear whether any of his early writing was left at CMC and survived.*

Fig. 1. D.R.P. Coats as a Marine Wireless Operator in 1915. (from *Manitoba Calling*, Vol. III, No. 10, 1939)

came across a draft history at CMC, and this was to have been the starting point for a book on CMC history. The book was never published, however, and in 1995 Roger sent a draft of his manuscript to Robert Murray. This article is the product of that draft and subsequent discussions.

The Early History

By January 1901, Guglielmo Marconi had succeeded in establishing wireless communication over a distance of 200 miles, a great advance over the 1-1/2 miles that he had first achieved shortly after his arrival in England five years earlier. Despite scientific opinion to the contrary, Marconi was convinced that the curvature of the earth would not prove to be a barrier to the transmission of radio signals over longer distances, and he decided to attempt to establish two-way wireless communication across the Atlantic Ocean. As a result, work was started in 1900 on stations at Poldhu in Cornwall, England, and at Cape Cod in Massachussets, and testing began in early 1901.

As is often the case, fate, rather than planning, played a large part in the location of the receiving station for the first trans-Atlantic wireless transmission. In September 1901, a severe gale wrecked the massive Poldhu aerial system and at about the same time, another storm wrecked the antenna system at Cape Cod. A simpler antenna system was quickly erected at Poldhu, but Marconi decided not to wait for completion of the reconstruction at Cape Cod, as he feared that rivals such as deForest and Fessenden might gain the lead. Marconi therefore modified his original objective, and decided to attempt only a one-way transmission across the Atlantic, from the transmitter in Poldhu to a receiving site he was to establish in Newfoundland. On September 17, Marconi and his assistants, Kemp and Paget, duly set sail for Newfoundland.

Marconi's hypothesis about the long-distance transmission of wireless waves was finally proved to be correct when, on December 12th, 1901, at the site near St John's, Newfoundland, he received the letter "S" in Morse code transmitted from Poldhu.

Spurred on by the success of the Poldhu to Newfoundland tests, Marconi decided to set up a permanent sending and receiving station in Newfoundland. Unfortunately, on December 16, 1901, he was informed that his activities constituted an infringement upon Anglo-American Telegraph Company rights in Newfoundland and was threatened with legal proceedings. He therefore decided to return to England to await the completion of the

Fig. 2. Marconi (centre), Kemp and Paget on the steps of the tower at Signal Hill, St. Johns, Newfoundland, on December 12, 1901. Public Archives of Canada.

station on Cape Cod. However, four days later, on December 20, 1901, the Canadian Minister of Finance, W.S. Fielding, wrote to Marconi expressing his pleasure at the suggestion that the experiments might be continued in Nova Scotia. Fielding assured him of a cordial welcome and the co-operation of Government officials and of the absence of difficulties that might get in the way of his operations.

When Marconi arrived at North Sydney in Cape Breton, Nova Scotia, on December 26, he was met by a reception committee that included the Hon. G.H. Murray, Premier of Nova Scotia; Cornelius Shields, General Manager of the Dominion Coal Company; and Alex Johnston, owner and editor of the *Sydney Record* and a local Member of Parliament. As a reporter, Johnston was naturally interested in meeting Marconi, but he was also there for another reason. Johnston realized the tremendous potential of wireless communication and, being aware of the legal difficulties Marconi had encountered in Newfoundland, he intended to persuade the inventor that Cape Breton would be equally suitable for his purpose. While Marconi was in North Sydney, Johnston managed to convince him that he had a site that was worth serious consideration, and Shields arranged for a special train to take Marconi, Kemp and Paget to Table Head near Glace Bay. Despite several other offers of land, (including one from Alexander Graham Bell), Marconi quickly decided that the Glace Bay site was the most suitable, and set about raising the capital to build a trans-Atlantic wireless station. In this endeavour he was again greatly helped by Johnston and by A.S. Kendall, another Member of Parliament, who accompanied him to Ottawa to discuss the project with Johnston's personal friend and political colleague, the Hon. W.S. Fielding, Minister of Finance. Through Fielding, the group was able to confer with the Prime Minister, Sir Wilfred Laurier, and the Postmaster General, Sir William Mulock. As a result, an agreement was entered into on March 17, 1902 between Marconi's Wireless Telegraph Company Limited (MWT), the Marconi International Marine Communications Company Limited (MIMCo), and His Majesty, King Edward VII, represented by the Prime Minister, the Rt. Hon. Sir Wilfred Laurier, President of the King's Privy Council for Canada.

By this Agreement, MWT and MIMCo were to erect two wireless stations, one in the United Kingdom and the other in Nova Scotia, with the object of carrying on commercial communications between Canada and the United Kingdom and Europe. The Canadian Government undertook to defray the costs of the erection of the Canadian station up to a limit of approximately Cdn$80,000. Any excess costs were to be borne by MWT and MIMCo. The companies also agreed that if the undertaking was successful the charge for ordinary messages would be not more than 10 cents per word (versus the 25 cents per word then charged for cables), with Government and Press messages to be handled for only 5 cents per word.

In addition, the Agreement gave the Canadian Government the right to erect stations as it saw fit, using the Marconi system of wireless telegraphy "for communication with any of its lighthouses or life-saving stations on the Coast, or between the Mainland and any island within the jurisdiction of Canada, or with any ships passing to and fro, or in any way to assist in its operations for the protection of life and property on the seacoast or inland waters of Canada, or for the improvement or assistance of navigation". For their part, the companies agreed to furnish all equipment necessary at "fair and reasonable prices free from any charge for patent rights or royalties thereon". In addition, the companies acquired the option, whenever the Government expressed its intention to erect a wireless station, of installing and operating it themselves and retaining for their own use "all tolls collected at such stations for messages transmitted therefrom to passing ships". Marconi and his associates saw this Agreement with the Canadian Government as the essential ingredient in ensuring the successful and profitable promotion of a new company in Canada. Accordingly, on July 30, 1902, Marconi's Wireless Telegraph Company entered into an agreement with Willard Reed Green of New York, acting both for himself and as agent for a Canadian company he was to form. Within 14 days after the practicability of reception in Canada of the transmission of messages transmitted by the Marconi system from England had been established, Green was to make application for the incorporation of a new company to be called "The Marconi Wireless Telegraph Company of Canada" (MWTC). The company was registered as of November 1, 1902.

The new company was to have a nominal capital of $4,999,500, in the form of 999,900 shares of $5 each, and five Directors, three to be nominated by Marconi's Wireless Telegraph Company and two by Green. For its part, MWT agreed to sell to MWTC:

1. all rights existing in Canada and Newfoundland to all inventions in wireless telegraphy and all letters patent thereon which MWT then owned or might at any time thereafter possess;
2. the benefit of all contracts entered into by MWT connected with the use of the Marconi system in Canada and Newfoundland; and
3. all other property, rights, privileges, etc., of MWT in that territory.

In return, the new company was to allot to MWT 660,000 fully paid shares at a par value of $5 each. One hundred thousand shares were allotted to Green for his services, and he also received an allotment of 40,000 shares to dispose of as he saw fit to "defray the costs of advertising and his expenses in selling shares". The remaining 200,000 shares were allocated to provide working capital for MWTC. Green had the sole privilege of selling up to 440,000 shares at any time up to three months after the incorporation of MWTC, 100,000 of which were to be provided by Marconi's Wireless Telegraph Company, 200,000 by MWTC, and 140,000 by himself. The only restriction placed upon Green was that the shares should not be sold at less than $2.50; it appears that any amounts received in excess of that price would belong to him.

The first test transmission from Glace Bay was made on November 19, 1902, but reception at Poldhu was poor. Several modifications to the equipment and antennas were made over the next two weeks, and on December 5 the receiving station at Poldhu reported the first readable signals from West to East.

Reception continued to be sporadic, however, and it was not until December 21 that Marconi sent a formal message to England from the new station, saying that he had established trans-Atlantic wireless telegraphic communication "with complete success".

On January 3rd, 1903, the first CMC Directors' meeting was held, attended by J.N. Greenshields, W. Barwick, H.C. Osborne, J. Payne, and Willard R. Green, who became the first president of the Company. At the meeting, ByLaw 23 was passed, authorising the stock issue and arrangements with MWT. This agreement was to govern the relationship between the two companies for nearly 50 years. It re-stated the relationship between the Canadian assets of MWT that were to be transferred to CMC, in accordance with the original agreement with Willard Green, and then went on to make further provisions, including:

1. Marconi's International Marine Company (MIMCo), which by an agreement with MWT dated April 30, 1902, had acquired all rights to the Marconi system in the marine field, assigned to CMC all its rights in Canada and Newfoundland.
2. MWT and MIMCo transferred to CMC the benefit of the contract of March 17, 1902 with the Canadian Government concerning the establishment of a station at Table Head.
3. All gross tolls from communications between the United Kingdom and Canada were to be divided equally between MWT and CMC.
4. All tolls for communications between any vessels at sea and any shore stations, or between two or more vessels at sea, were to accrue to the benefit of the company owning the rights for the country to which the vessel or shore station despatching the message belonged.
5. If a Canadian ship were fitted in the U.K. by MIMCo with the Marconi system, half the net profit from the manufacture and installation of the equipment was to be paid to CMC. There was a similar arrangement regarding U.K. ships fitted by CMC.
6. If CMC acquired any invention in the field of wireless telegraphy, it would assign all rights thereunder, other than the Canadian ones, to MWT.

It should be noted that this agreement implies that the principal, and perhaps only, sources of revenue contemplated for CMC were trans-Atlantic and marine communications; there is no evidence that anyone at that time envisaged any other use for wireless telegraphy. It is also worth noting that the physical assets transferred by MWT to CMC under the 1903 agreement were nominally valued at $100,000, so that the payment to MWT for patents and other rights amounted to $3,200,000 in the form of stock and seem, therefore, to have been regarded as the principal asset of the new company. However, although these were a wasting asset, no action was taken to write them down on the books until 1919.

Over the next few days, Mssrs. Payne, Osborne, Barwick and Greenshields resigned, and were replaced by Mssrs. J.D.Oppe, A. Allen, R. Forget, and Col. F.C. Henshaw. In addition, J.D. Oppe was appointed General Manager. At a General Meeting of Shareholders on January 10, 1903, 900,000 shares were represented, as follows: 560,000 MWT, England; 339,920 W.R. Green; and the balance by the Directors. At the meeting, Guglielmo Marconi was appointed Vice-President, and his salary established at $2,500 per annum. Two days later, the Company's offices were established at 1724 Notre Dame Street East in Montreal.

For a number of years, MWT had been in a dominant position in wireless telegraphy and hoped to establish a world-wide monopoly. Marconi's first patent had been issued in 1896 and he had obtained a number of others during the next three or four years. All of these now seem very primitive indeed, for in none of them is there any evidence that Marconi was aware of the principle of resonance, which is fundamental to the whole radio art. In fact it was not until the publication of his famous patent number 7777 in April 1900 that the use of this principle was disclosed. This patent, however, greatly strengthened Marconi's position and must have encouraged him and his associates in their ambitions for world monopoly. They had set up the American Marconi Company in 1899 and other subsidiaries in Belgium, France, Russia, Spain, and several other countries before 1910. They had acquired by contract with Lloyd's the right to set up wireless stations, for communications with ships, in any of the signal stations of Lloyd's world-wide network and pursued a vigorous policy of constructing coastal stations, at the same time that they were equipping ships. During the early years they discouraged competition, particularly from Germany, by forbidding their coast stations to communicate with ships not equipped with Marconi equipment. The first real obstacle in this march towards monopoly occurred in 1906 when the Berlin Convention was ratified, requiring free interchange of wireless messages irrespective of the systems employed.

The Canadian Marconi Company

By a special act of Parliament, The Marconi Wireless Telegraph Company of Canada, Limited was granted a Dominion Charter on August 13, 1903, without any change in name. The books were closed for the first time as at January 31st, 1904, showing a deficit of $33,125. This represented expenses incurred at Montreal and Glace Bay, as no revenue was earned during that period. Annual financial results from this time forward until 1947 are shown below in Table 1.

Table 1. Canadian Marconi Company - Financial Statements

Cal Year	Fisc Yr End	Net Profit (Loss)	"Net Worth"	Capital Stock No. Auth	Par Value	Total Value Auth	Issued
1903	Jan 31 1904	($33,125.26)	$5,030,292.87	5000000	$1.00	$5,000,000	$5,000,000
1904	Jan 31 1905	($36,199.26)	$5,137,256.63	5000000	$1.00	$5,000,000	$5,000,000
1905	Jan 31 1906	($3,195.69)	$5,274,122.70	5000000	$1.00	$5,000,000	$5,000,000
1906	Jan 31 1907	($55,583.23)	$5,398,260.76	5000000	$1.00	$5,000,000	$5,000,000
1907	Jan 31 1908	($117,023.86)	$5,536,968.05	5000000	$1.00	$5,000,000	$5,000,000
1908	Jan 31 1909	($73,968.75)	$5,622,824.48	5000000	$1.00	$5,000,000	$5,000,000
1909	Jan 31 1910	($42,833.38)	$5,733,183.60	5000000	$1.00	$5,000,000	$5,000,000
1910	Jan 31 1911	($45,06.61)	$5,838,272.94	5000000	$1.00	$5,000,000	$5,000,000
1911	Jan 31 1912	($5,467.33)	$5,441,681.00	5000000	$1.00	$5,000,000	$5,000,000
1912	Jan 31 1913	$62.11	$5,573,959.50	5000000	$1.00	$5,000,000	$5,000,000
1913	Jan 31 1914	$7,158.62	$5,707,753.95	5000000	$1.00	$5,000,000	$5,000,000
1914	Jan 31 1915	$50,020.11	$5,743,866.63	5000000	$1.00	$5,000,000	$5,000,000
1915	Jan 31 1916	$110,226.06	$5,808,083.63	5000000	$1.00	$5,000,000	$5,000,000
1916	Jan 31 1917	$104,929.45	?	5000000	$1.00	$5,000,000	$5,000,000
1917	Dec 31 1917	$114,953.09	$5,947,019.92	5000000	$1.00	$5,000,000	$5,000,000
1918	Dec 31 1918	$138,647.98	$6,024,336.94	5000000	$1.00	$5,000,000	$5,000,000
1919	Dec 31 1919	$102,130.84	$3,817,621.17	1500000	$2.50	$3,750,000	$2,500,000
1920	Dec 31 1920	$69,906.38	$6,394,727.36	3000000	$2.50	$7,500,000	$5,500,000
1921	Dec 31 1921	$12,253.31	$6,424,737.64	3000000	$2.50	$7,500,000	$5,525,000
1922	Dec 31 1922	$22,251.36	$6,841,044.31	3000000	$2.50	$7,500,000	$6,000,000
1923	Dec 31 1923	($39,144.73)	$7,112,959.59	3000000	$2.50	$7,500,000	$6,300,000
1924	Dec 31 1924	($42,510.70)	$3,579,939.51	7500000	$1.00	$7,500,000	$2,520,000
1925	Dec 31 1925	$37,984.20	$4,224,876.23	7500000	$1.00	$7,500,000	$3,380,642
1926	Dec 31 1926	$24,362.71	$4,338,089.59	7500000	$1.00	$7,500,000	$3,380,642
1927	Dec 31 1927	$46,555.55	$4,888,692.34	7500000	$1.00	$7,500,000	$4,504,682
1928	Dec 31 1928	$181,590.97	$5,062,135.76	7500000	$1.00	$7,500,000	$4,554,682
1929	Dec 31 1929	$162,806.53	$5,212,987.28	7500000	$1.00	$7,500,000	$4,554,682
1930	Dec 31 1930	$13,441.00	$5,281,582.52	7500000	$1.00	$7,500,000	$4,554,682
1931	Dec 31 1931	($98,651.29)	$5,148,492.24	7500000	$1.00	$7,500,000	$4,554,682
1932	Dec 31 1932	($188,085.96)	$5,083,972.80	7500000	$1.00	$7,500,000	$4,554,682
1933	Dec 31 1933	($173,523.98)	$4,955,957.14	7500000	$1.00	$7,500,000	$4,554,682
1934	Dec 31 1934	$60,386.78	$4,794,809.69	7500000	$1.00	$7,500,000	$4,554,682
1935	Dec 31 1935	$96,159.98	$4,830,929.76	7500000	$1.00	$7,500,000	$4,554,682
1936	Dec 31 1936	$189,316.15	$5,004,606.21	7500000	$1.00	$7,500,000	$4,554,682
1937	Dec 31 1937	$268,376.96	$5,379,282.56	7500000	$1.00	$7,500,000	$4,554,682
1938	Dec 31 1938	$335,914.92	$5,508,104.49	7500000	$1.00	$7,500,000	$4,554,682
1939	Dec 31 1939	$278,030.17	$5,815,337.24	7500000	$1.00	$7,500,000	$4,554,682
1940	Dec 31 1940	$493,763.06	$6,229,730.70	7500000	$1.00	$7,500,000	$4,554,682
1941	Dec 31 1941	$572,209.40	$7,602,464.37	7500000	$1.00	$7,500,000	$4,554,682
1942	Dec 31 1942	$329,959.35	$10,964,593.04	7500000	$1.00	$7,500,000	$4,554,682
1943	Dec 31 1943	$180,639.02	$12,276,602.58	7500000	$1.00	$7,500,000	$4,554,682
1944	Dec 31 1944	$198,709.66	$10,938,781.54	7500000	$1.00	$7,500,000	$4,554,682
1945	Dec 31 1945	$197,605.26	$8,389,846.99	7500000	$1.00	$7,500,000	$4,554,682
1946	Dec 31 1946	($25,947.41)	$8,459,697.10	7500000	$1.00	$7,500,000	$4,554,682
1947	Dec 31 1947	$67,872.12	$9,105,812.73	7500000	$1.00	$7,500,000	$4,554,682

from Marconi Wireless Telegraph Company of Canada Annual Reports

The Beginning of the Commercial Era

On May 13th, 1904, the first contract for construction and operation of coast stations was entered into with the Canadian Government. This called for stations to be erected at Belle Isle, Point Amour, Heath Point, Fame Point, and Cape Race, these stations to be owned by the Canadian Government but operated and maintained by CMC. In addition, CMC agreed to erect, at its own expense, a station on Sable Island and one on the mainland; on August 12, Cape Ray was added to the Government stations.

In an attempt to produce more consistent reception, the power of the Glace Bay station was increased to 150 kilowatts and the wavelength was changed to 3660 meters. In addition, 24 lattice towers were erected, (at a cost of over $62,000) to support a "square cone" of 400 copper wires, so constructed that all or part of the complete system could be utilised at any one time (Figures 3, 4 and 5). Great difficulties were still experienced, however, in establishing a reliable transatlantic radio service, and it was soon realised that mere increases in power and antenna size would not overcome these problems. Attempts to establish the service were therefore curtailed until the technology was equal to the task. A new site for the Glace Bay station was purchased for $7,386 at Port Morien, six miles from the original site at Table Head, as the latter could not accommodate the larger antenna system required for trans-Atlantic communication. This new site (still referred to as Glace Bay) had an area of over 600 acres, or approximately one square mile, allowing the construction of an antenna consisting of 200 wires stretched out in an umbrella shape and anchored to supports forming a circle with a diameter of almost half a mile. By 1906, the station had cost some $150,000; since revenues for the year were a mere $150, these funds were borrowed from MWT in England.

Fig. 3.   Erecting lattice towers at Glace Bay

Fig. 4.   Glace Bay tower construction

Fig. 5.   Marconi Wireless Station, Glace Bay, before the move to Port Morien.

During 1905, additional coast stations were constructed, operated and maintained on behalf of the Canadian Government. Stations were built by the Company at Sable Island and Halifax at a total cost of approximately $15,500. There was also a temporary (summer) station at Whittle Rocks on the north coast of the St Lawrence River, which formed part of a chain to Flame Point. Ships tolls revenues for the year were some $1,403.55.

During the period 1906-1908, the history of the Company is one of continual improvement of the facilities at Glace Bay and the construction of additional stations on the East Coast on behalf of the Canadian Government. An increasing number of vessels were being fitted with the Marconi system at this time, many of which were in communication with the stations that had been established on the East Coast. Ships tolls revenues for 1906 were $7,339.82.

In 1907, a new station was built by MWT at Clifden in Ireland, operating on a wavelength of 6666 meters, and by February, 1908 an unlimited service between Clifden and Glace Bay had been established (Figure 6). At year end, the Company owned stations at Cape

The Marconi Company in Canada, 1906

When I arrived at the station in Glace Bay I was told that the manager, Mr. Vyvyan, had not yet arrived and I was advised to proceed down the drive to his residence. About half way I met a young man in tennis flannels and asked him if I was going in the right direction to see Mr. Vyvyan. He told me that Mr. Vyvyan was on his way to the station and I would meet him on his way. I also asked if there was any likelihood of a vacancy at the station. He was most polite and helpful and then went on his way. In about fifteen minutes I met Mr. Vyvyan and told him the reason for my visit. Mr Vyvyan surprised me by saying, "I see you have already met our Mr. Marconi." I was really staggered to hear this and on giving Mr. Vyvyan a brief outline of my studies and working life to date he promised to send me notice as soon as he had a suitable vacancy. Several months later I received a message requiring me to report at the station. This I did and was employed to assist Mr. A. Warren in the machine shop. This was a few weeks before the news was received at the station of the death of Edward VII in England.

From unpublished recollections of William Appleton, MBE, former CMC employee.

Fig. 6. Glace Bay, October 1907, L.R. Johnston is the operator.

Sable, North Sydney, Pictou, Three Rivers, Sable Island, and Halifax, in addition to the Glace Bay station. For the first time, Glace Bay is shown as making a contribution to the income side of the ledger, with revenues for the year of $2,012 for trans-Atlantic traffic and $433 from manufacturing. In 1908, these revenues rose to a total of $7,875, including $5,846 for trans-Atlantic tolls and $1,700 for ship tolls.

In 1909, CMC opened its first Montreal workshop, on Delorimier Street, for the manufacture of apparatus used on low-power stations and as a repair depot (Figure 7). In August, the operating house at Glace Bay was destroyed by fire, necessitating the cancellation of the trans-Atlantic service until April 23, 1910. However, this disaster gave the Company an opportunity to re-design the station building and to introduce modern equipment for long-distance wireless telegraphy, including the replacement of the alternating current supply with a direct current supply derived from a massive storage battery and a huge air condenser.

By the end of the year, CMC was operating 26 coast stations, 15 of which were owned by the Canadian Government, five by the Newfoundland Government, and the remaining six by the Company. On all Government-owned stations, CMC received a yearly subsidy and also retained the traffic tolls, which amounted to $36,100 for the year ended January 31, 1910.

By the end of 1910, 29 stations in Canada and Newfoundland were owned or controlled by CMC, including the one at Camperdown, Nova Scotia, pictured here (Figure 8). At year end, the Marconi system had been installed on 536 vessels, of which 223 were large passenger steamers regularly plying the North Atlantic route and communicating with stations controlled by the Company "for up to three days before arriving at and after leaving port". Traffic receipts for fiscal 1910 amounted to $42,630.

The development of the Marconi System through the Great Lakes was started in 1910 with the establishment of a station at Port Arthur, Ontario, and the equipping of three vessels.

On February 21, 1911 a sweeping decision entirely in favour of MWT was given in the High Court of Justice, London, England. In the case of MWT versus the British Radio Telegraph and Telephone Company Limited, the latter were adjudged infringers upon Marconi's famous patent 7777, which covered the principle of resonance, essential to the efficient operation of any wireless telegraph station. This decision was very important, as it greatly strengthened the position of the Marconi companies throughout the

Fig. 7. CMC first factory on Delorimier St. in Montreal

One morning I was awakened by the sound of falling buildings and I saw that the transmitting house was fiercely ablaze. All hands including the staff were quickly on the spot but the water supply and extinguishers were unable to cope. In the spark room were several tanks containing hundreds of gallons of insulating oil, in addition to oil filled transformers. All that was left of the station was the power house and machine shop, which were about 200 yards from the fire. Practically nothing was salvaged.

After this there was nothing to be done except get busy with ordering new machines and making new transmitting apparatus. The engineering staff together with Warren and myself were kept busy almost 20 hours a day seven days a week for several months. I often managed to get a couple of hours sleep upon a 20" wide engine belt in the power house. A new system of producing high frequency energy was adopted which included 6,600 large accumulators in series which were charged from a group of 5,000 volt DC generators at 15 amps initial rate.

It was on one Good Friday that forty labourers plus all the staff including Mr. Marconi himself were assembled early to begin the filling of the cells. Every available acid proof utensil was commandeered and work commenced. About 20,000 gallons of acid were used up and many were spilled over our clothing and shoes by the end of the day. Most of the clothing had to be scrapped. During these operations all suffered intensely from the irritation caused by the acid and the fumes. The language of most would have caused another fire but Mr. Marconi was able and did express his feelings in his native Italian.

From unpublished recollections of William Appleton, MBE, former CMC employee.

world and enabled them to retain a virtual monopoly on wireless telegraphy.

Another important event in 1911 was the signing of an agreement, dated April 5th, between the Company and the Canadian Government, which superseded and replaced all previous agreements and contracts and established the basis of operation for all Coast Stations operated by CMC. This agreement provided that CMC would be paid specified fixed subsidies, and in addition, would be entitled to retain all Coastal Station traffic tolls. The contract also provided that, in consideration of the payment which had been made by the Government under the 1902 agreement with MWT, CMC would not charge more than 10 cents per word for commercial traffic and 5 cents per word for Press traffic handled through the Glace Bay station. In addition, no charge was to be made to the Government for messages handled through any of the Coast Stations operated by CMC. The contract was to remain in force for 20 years and provided for payment of full subsidies for the first 10 years and for a reduction in subsidies at the rate of 10% per annum over the second 10 years.

By the end of 1911, the Company was involved in the operation of some 33 shore stations. It was in 1911 that the plan to erect a chain of long-distance wireless telegraph stations "to girdle the earth" was first announced. Under a contract between the British Government and Marconi's Wireless Telegraph Company, the latter was to receive $300,000 for each station, exclusive of sites, foundations, and buildings, plus 10% of the gross receipts for all stations for a term of 28 years. As it turned out, however, it was to be a considerable period of time before the plan was actually realized. Also in 1911, the Marconi Wireless Telegraph Company of America absorbed the United Wireless Telegraph Company, which had been a fierce competitor in certain areas. A number of steamers of Canadian register, formerly operated by United, consequently came under the control of the Company.

Over the previous few years, the trans-Atlantic business of the Company had been subject to very strong competition from the land telegraph companies of Canada and the US, all of which were affiliated with various cable companies. An agreement with the Western Union and Great North Western Telegraph companies, signed in 1911, provided for their handling of trans-Atlantic Marconigrams, and for the acceptance and delivery by the telegraph companies lines, offices, and messengers on the

Fig. 8. Camperdown, Nova Scotia, Marconi Wireless Station

same terms as those enjoyed by the cable companies. This agreement, together with an anticipated increase in capacity, was expected to greatly augment revenues from trans-Atlantic traffic.

Early in 1911, the Company, to some extent as a result of Government urging, announced its intention to establish a chain of stations along the Great Lakes to link the exisitng stations at Montreal and Port Arthur. A draft agreement was prepared, under which the Company was to equip and operate four stations, receiving from the Government the sum of $3,500 per year per station, in addition to retaining 85% of the traffic receipts. A further five stations were identified as being required to complete the

Fig. 9. Operating room, Marconi Station at Port Edward, Ontario

Fig. 10. Marconi Station at Midland, Ontario, showing switchboard in foreground

chain, and these were to be operated under the same terms as they were completed. On September 7th, 1912, the agreement was signed under which the Company agreed to maintain and operate eight stations on the Great Lakes on behalf of the Canadian Government, including those at Midland and Point Edward shown here in Figures 9 and 10.

At the end of the year, the Company had contracts for the installation of Marconi apparatus on a total of 44 vessels, including 12 Great Lakes vessels, 10 steamers on the Atlantic coast, 11 on the Pacific coast under Company control, three trans-Pacific steamers controlled on behalf of the Marconi International Marine Co., and a further four operated on behalf of affiliated companies. A standard contract at that time was for a period of five years. A small CMC marine set of the time is shown in Figures 11 and 12.

Traffic receipts for ship-to-shore and inter-station business continued to increase, and the total for the year amounted to $52,322. The capacity of the Glace Bay station was increased, and it was decided to erect a duplex receiving station at that location. The total receipts for the year for trans-Atlantic traffic amounted to $44,950.

The contract providing for the erection of a chain of long-distance wireless telegraph stations to girdle the earth was ratified in 1913, sites were selected, and construction began. Then came World War I, with strict Government control of all communications, and work on the project was suspended.

Early on Sunday morning, April 14th 1912, normal communication was established by the CMC station

Fig. 12.   ¼ Kw marine set, side view

Fig. 11.   ¼ Kw marine set, front view

Fig. 13.   Condenser house, Glace Bay, May, 1912

Fig. 14. Operating table, Glace Bay

at Cape Race with the S.S. Titanic, which was then 800 miles to the east. All was normal until 12:25 a.m. on April 15th, when a distress signal was received from the Titanic indicating that she had struck an iceberg and was in serious danger. Messages were immediately sent from Cape Race to other ships within range, several of which indicated that they were altering course and proceeding to the scene. The station also telegraphed the ships owners at New York, Liverpool, and Southampton and within half an hour telegrams began pouring in from many distant points requesting more information. Indeed, Cape Race was the only Coast Station that handled ship-to-shore communications in the early stages of the disaster. This disaster brought to the attention of the world the great advantage of wireless communication and had a marked impetus on its development over the next few years.

The CMC annual report for the period ended January 31, 1914 states that the business done under contract with the Canadian and Newfoundland Governments had expanded "satisfactorily", but that ship traffic receipts had been seriously interfered with by the destruction of the operating house at the Cape Ray Station by fire on May 5. Despite the establishment of a temporary station within two days, it was not until September 30 that a full commercial service was resumed. Inside views of the Glace Bay station of the period are shown in Figures 13 and 14.

Until 1913, both Glace Bay and Clifden trans-Atlantic stations had operated on a simplex basis, that is, neither station could send and receive at the same time. In the latter part of that year, duplex systems allowing simultaneous reception and transmission were instituted, but the proximity of the receiving and transmitting stations caused insurmountable problems due to interference. This was a serious drawback, as introduction of the duplex system would have doubled the capacity of each station, and thus allowed much greater revenues. By early 1914, both Clifden and Glace Bay had solved this problem by the simple expedient of moving their receiver sites 25 miles from the transmitter stations, the Clifden receiver station being located in Letterfrak, and the Glace Bay receiver station at Loiusbourg (Figure 15). The extra space also permitted the installation of directive aerials that gave optimum reception in the direction of the distant station and reduced interference from the local station to a minimum.

In November 1913, a great blizzard occurred and the eight coastal stations then established on the Great Lakes, including the Sarnia station that had been opened in February of that year, played a major role in providing assistance to navigation. In recog-

Fig. 15. Marconi Receiving Station at Louisbourg, Nova Scotia

Fig. 16. Marconi Station, Port Nelson, Hudson's Bay

Fig. 17. One of the 350' steel aerial masts at Port Nelson

Fig. 18. Receiving Room, Port Nelson

nition of the efforts of the CMC employees on these stations, several of the Officers-in-Charge received medals from United States ship owners.

Under a contract with the Department of Railways awarded in 1912, communication was established in 1913 between two new high-power stations at La Pas, Manitoba, and Hudson's Bay. However, completion of these stations was delayed due to lack of transportation facilities, and they were not finished until the following year. Figure 16 shows a station building at Port Nelson, Hudson's Bay, and Figure 17 a more detailed view of one of the 350 ft. steel towers. Figure 18 shows the receiving room.

As a result of the aftermath of the Titanic disaster in 1912, the Maritime nations of the world passed regulations requiring all ships of certain classes to carry wireless equipment. This led to a period of expansion, and by 1913 MIMCo had secured or was negotiating for concessions in more than 20 foreign countries. In addition, Government statistics for 1912 show that fully 90 of a total of 93 vessels of Canadian registry were equipped with the Marconi system.

World War I brought all normal commercial wireless development to a halt, but it also greatly stimulated the development of the vacuum tube, whose use and further development after the war was to transform the world of wireless into the world of radio. It seems, however, from the events that occurred shortly after the war, that the minds of the leading men in the wireless business were still pre-occupied with its principal function of trans-oceanic and maritime communications.

The war also created a demand for new ships, and large numbers of vessels were constructed in Canada for both the British and Canadian Governments. Government regulations required all these vessels to be equipped with wireless apparatus, with the result that the Company had to substantially increase pro-

Fig. 19. Works Department, Rodney St., Montreal, 1914, showing a Tuner type 843 at the left side in the foreground.

Fig. 20.   Marconi staff, Rodney St. Works, early 1915

duction in order to meet the higher demand.

There was a net profit of $50,020 for 1914, and a surplus of $5,728. In addition, claims were submitted to the naval authorities to compensate for the loss of revenues from the coastal stations. Branch offices and stores were opened in Toronto and Vancouver, and a well-equipped, three-story factory with excellent shipping facilities was established in Montreal on Rodney Street (Figures 19 and 20). The office facilities were moved at about the same time to Room 507, Shaughnessy Building, 137 McGill Street.

Coastal stations at Port Burwell, Toronto, and Kingston were formally commissioned in February, 1914. The Kingston station was fairly typical of these, having a "standard" 5.5 kilowatt disc transmitter built by CMC, an earthing network of some 2000 feet of copper netting, and three-section antenna masts 185 feet high. The system had a nominal wavelength of 600 meters, but transmissions varied from 600 to 1600 meters, and reception from 300 to 3000 meters. Receiving equipment included both a Marconi "Maggie" magnetic detector and a carborundum crystal detector (Figures 21, 22 and 23). At the Cape Race station, the 160ft wooden spar masts were replaced with 250 ft steel ones, greatly increasing the range of the station.

During the war years, traffic to and from ships was adversely affected by the severe censorship

Fig. 21.   Transmitting Room, Marconi Station, Kingston, Ontario

imposed, the general dislocation of passenger traffic, and the placing of important stations at the disposal of the Government. There were, however, important gains in the volume of trans-Atlantic traffic, and the over-all result of operations for these years showed considerable improvement, profits being earned each year from 1914 to 1918 inclusive.

In 1915, a school of instruction for operators was established in Montreal under the direct control of the Company, and equipped with the latest apparatus (Figure 24). In this year also, a Marine Depot was established in St. John's, Newfoundland.

Fig. 22. Operating Table, Marconi Station, Kingston

Fig. 23. Close-up of Tuner type 843, Marconi Station, Kingston

Fig. 24. Darby Coats instructing at the school for operators in Montreal

Fig. 25. William St. Factory. Experimental broadcast station XWA was later installed in the top left corner.

In 1917, there was an increase in the number of vessels of the Canadian Mercantile Marine equipped with Marconi apparatus, which increased the revenue from rental contracts and also from ship tolls. During this year, also, the Company purchased the building that it partially occupied at 173 William Street, Montreal, and equipped it with modern apparatus (Figure 25). Later, the factory on Rodney Street was closed and all operations were transferred to the William Street premises.

On August 4th, 1917, the Company's trans-Atlantic stations at Glace Bay and Loiusbourg were closed to public service by order of the British Admiralty. Claims were rendered against the British Government for loss of revenue incurred by this suspension of service. In the same year, a basis of settlement for the two-year period ending 31 July, 1917, was arrived at with the Canadian Government in compensation for the closing down of certain East Coast stations. Following the signing of the armistice in 1918, these restrictions were gradually lifted and, as of December 2, 1918, the stations were allowed to re-open and handle Press and Government traffic; authority from the Admiralty to accept commercial messages, however, was not received until March 10, 1919.

> *I first became interested in radio only after I had been given a job by Marconi's Wireless Telegraph Company of Canada Limited, which began on January 13, 1919. It has to be appreciated that at that time there was no radio broadcasting, and hence amateur work was limited to a relatively small group who were concerned with radio telegraphy. However, very soon after I joined the company, broadcasting commenced and my interest and that of most other technically minded people began at that stage; confined of course to the various forms of crystal receiving sets at the beginning.*
>
> *I joined the company as an engineering apprentice for a 5 year term. The idea being that during that term the apprentice would be trained in very nearly all, if not all, of the technical sides of the business. I started in the shop and spent the better part of two years there. During that time we were given courses in radio telegraph operating, in rigging of masts and aerials, splicing of ropes and a number of other things of that kind. Later we were introduced to test work now called quality control, and of course there was a very great deal of installation work to be done on ships and on coast stations. I applied to the company for a leave of absence to attend university beginning in the Fall of 1920. The university program and the apprenticeship worked very well together, and thus I spent four Winters at the university and was more or less guaranteed a summer job. I know of no other apprentices in this company who at the same time studied engineering at the university.*
>
> *I might say that at the very beginning the rate for an apprentice was $30.00 a month with no fringe benefits except a two week vacation and a 60 hour week. Later this was changed to a 55 hour week, and we felt a very great sense of progress indeed having Saturday afternoon off.*
>
> **Unpublished recollections of S.M. Finlayson, CMC President from 1951 to 1964.**

During 1917, 122 vessels were fitted, despite difficulty in obtaining some parts, notably motor-generators which had previously been imported and which the Company then decided to manufacture itself.

The Broadcasting Era

CMC's earliest experiments with radio-telephony were conducted by Mr. Arthur Runciman in 1919. A 500-volt battery was set up on a truck and a "Captain Round" type transmitter tube was connected to an antenna on the roof (see Tyne, G.F.J. *Saga of the Vacuum Tube*, 1977, Chapter 11.). Tests indicated a range of 3 or 4 miles, but in later experiments, with the transmitter set up on Tarte Pier, wireless operator Harris aboard the ice-breaker "Lady Grey" 30 miles down river from Montreal reported very good reception.

These tests were preliminary to the installation of a 500-watt transmitter imported from the MWT factory in the United Kingdom, and designed to permit continuous-wave wireless telegraphy, buzzer-modulated wireless telegraphy, or radio-telegraphy (Figure 26). Using this equipment, Canadian Marconi initiated broadcasting of the human voice and music in Canada when test programs were carried out by the Company from its premises on William Street in Montreal on a wave length of 1200 meters, using the call sign XWA. The first programs were quite amateur affairs, consisting of gramophone records, news items, and weather reports, with the announcing done by CMC employees. There was no schedule at this time, and no proper studio, the very first broadcasts being made from a laboratory. It is difficult to say just how large the audiences were for these transmissions, though they certainly grew rapidly as the new 'fad' of radio caught on. Certainly the first listeners were radio amateurs rather than the general public, for the latter had not yet been seized by the new medium.

Much has been written about which station should receive credit for being the one to initiate broadcasting in North America, with KDKA in Pittsburgh usually being given the distinction by

Fig. 26. 500-watt cabinet set imported from MWT factory at Chelmsford, England, and used at the experimental station XWA in Montreal. (*Manitoba Calling*, 1940, Vol. IV, No. 9, Page 8)

The first days of station XWA

Coats delivered a series of his reminiscences as broadcasts in Winnipeg, which were later reprinted in Manitoba Calling, the station bulletin. One such broadcast included the following:-

The transmitter had been shipped early in 1919 from Marconi's works in Clelmsford, Essex, England. In Montreal the radio studio was a whitewashed room on the top floor of a radio factory building on William Street. The radio set was in a vertical teak box that looked something like a piano. An engineer came up the stairs from the main floor where he had started a motor-generator which was to supply current to the wireless telephone. He entered this bare room which was the first Canadian radio studio and threw a switch. Three tubes lit up, not glowing dimly but shining with the brilliance of electric light. There was a pause of a few minutes to allow the tubes to become thoroughly warmed and ready for action. Then the engineer picked up the microphone, which looked much like a common telephone mouthpiece of that time. He held it close to his lips and spoke—thus: "Hello! Hello! This is wireless telephone station XWA at Montreal. Hello! Hello! How are you getting this? Is it clear? Is the modulation O.K.? XWA at Montreal is changing over".

The expression "changing over" meant he was ready to receive. There would have been maybe a few dozen people in Montreal at the time who were equipped to hear the transmission and understand it. The engineer was communicating with some other station operator with whom he was conducting his experiments. These first tests were on 1,200 metres (250 KHz.). Most radio traffic in 1919 was in c.w. so that the early days of voice transmissions were welcomed as a novelty. Soon the company inaugurated hour-long daily broadcasts, some of which were conducted by Coats. Some of the earliest broadcasts had been conducted by Jack Argyle, radio engineer, and J.O.G. Cann, chief engineer of CMC.

Radio programs began with the addition of music to speech at the microphone. To begin with, the terse sentences of the engineers, thrilling as they were to experimenters, had little to interest the public, to whom they were trying to sell receiving sets. The engineers too, ran out of breath and grew tired of repeating the alphabet and saying "ninety-nine". Probably personal convenience persuaded them to do less talking and fill the intervals while testing by playing phonograph records. In the interests of economy, the company refrained from buying a phonograph. Instead they asked the proprietor of a music store on Ste. Catherine West to lend them an instrument and records in return for suitable acknowledgements on the air. Thus the first "sponsored" programs from Canada went into the hitherto undefiled ether around Montreal.

Adapted from Coats, D.R.P. *Adventures in Radio* **– 14 & 15. In D.R.P. Coats (Ed.)** *Manitoba Calling*, **Vol. IV, Nos. 10 & 11, 1940.**

American sources and XWA being quoted by Canadian historians. The matter is unlikely to be solved definitively, however, since it hinges on a definition of what could be considered "broadcasting" at that time. Both the Canadian, Reginald Fessenden (in 1906), and an American, Lee de Forest made "broadcasts" that far predated any formal station. These are generally discounted, however, since they were experimental and unscheduled. By 1919-1920, both XWA and KDKA were making test transmissions, but it does appear that XWA was the first to broadcast regular, scheduled programs.

This commencement of broadcasting created a great deal of interest and opened up a new era in the field of advertising, although the full significance of this new medium was not realized at the time. It also created new fields of activity for CMC in the manufacture of both receiving and transmitting apparatus. Most of the early stations—including XWA—were owned by manufacturers of radio equipment, who developed test transmissions into scheduled programming so that purchasers of their products would have something to listen to! Later, when the idea of sponsorship—and therefore of advertising—caught on, it produced revenue that could be used for the production of better programs than individual companies could afford.

During the Fall of 1919, CMC decided to sell wireless apparatus to amateurs, and a separate company, Scientific Experimenter Limited, was formed for that purpose with headquarters at first at CMC headquarters at 11 St. Sacrament St., and then in 1922 at 33 McGill College Avenue in Montreal (Figure 27).

In parallel to their experimentation with broadcasting, CMC engineers were designing prototype receivers that were meant for the amateur market. One of these was a receiver using type V24 valves, made in England. The basic design was the same as the "C" set introduced later (in 1921), consisting of a detector and amplifier. A passive tuner would have preceded the detector (see Figures 28 and 29).

Another apparent prototype was the "Arcon Junior". The name was formed by leaving the letters M and I off the name Marconi. (For many years,

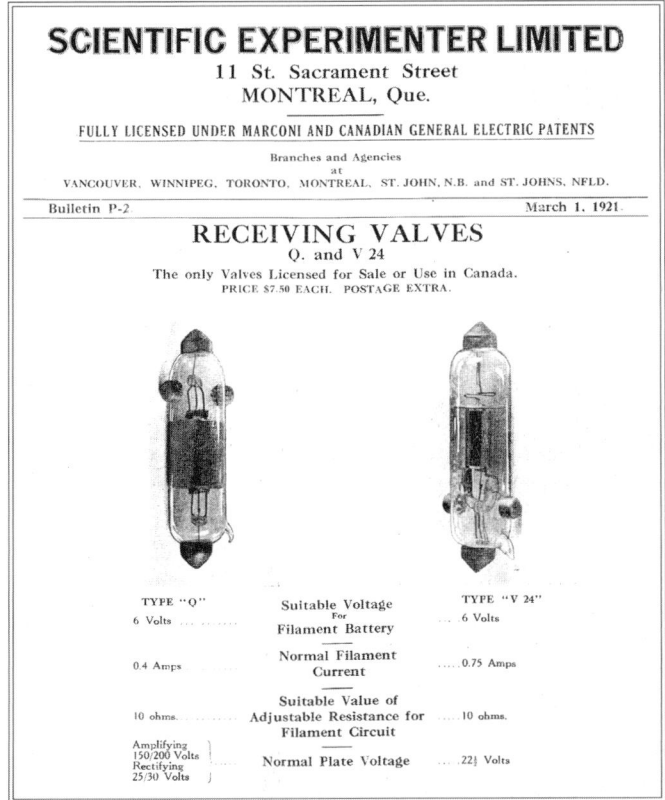

Fig. 27. Front page of Scientific Experimenter Limited catalog dated March 1, 1921.

Fig. 28. Early prototype detector using a Marconi V24 valve

Fig. 29. Early prototype amplifier using V24 valves

the cable code for the company was also "Arcon".) This receiver was possibly offered for sale, although only two examples are known to exist, one in the Canada Science and Technology Museum in Ottawa (Figure 30).

Still in the prototype stage of amateur receiver development, CMC produced the Type 55A consisting of 6 stages of radio-frequency amplification (V-24 valves) followed by a detector (Q valve), shown in Figure 31. The next photo shows a Type 55A amplifier and an unknown device, probably a receiver, the combination driving a Magnavox R-1 loudspeaker (Figure 32). The equipment is displayed in someone's garden, and the photo is dated 1920.

The first commercially produced amateur receiver was identified as the "C" set, and was first offered in 1921. It consisted of three separate boxes, a passive tuner, a detector, and a 2-tube amplifier. A separate power amplifier was also available (Figure 33). This set used the more readily available (in North America at least) UV-200 and UV-201 tubes. While in March of 1921, Scientific Experimenter Limited offered only the British Marconi Q and V24 receiving valves, by the following year they listed UV-200 and UV-201 receiving tubes and not Q and V24 types. They also listed UV-202 and UV-203 transmitting tubes.

Fig. 30. "Arcon Junior" prototype receiving set

Fig. 31. Type 55A amplifier

Fig. 32. Experimental combination of type 55A amplifier/detector, a presumed audio amplifier of unknown type, and a Magnavox model R-1 horn speaker.

Fig. 33. Marconi "C" set

Fig. 34. Marconi crystal receiver, catalog #139, Scientific Experimenter Limited

In 1922 they also offered two crystal sets, one a vertically mounted coil with 3 sliders and the detector on top (Figure 34), and the other a receiving set with variometer tuning that was intended to be coupled to the "C" set detector and amplifier (Figure 35). It also bears a striking resemblance to the AMRAD model 2575. As a result of objections from dealers and distributors, the operation of Scientific Experimenter Limited was discontinued in 1924.

Fig. 35. Marconi crystal receiver, catalog #138, Scientific Experimenter Limited (replica by Robert Murray).

At about this time, G. Marconi came to America to negotiate with General Electric for the exclusive world-wide rights to the new high-frequency alternator developed by Alexanderson. The negotiations were progressing favourably when the U.S. Navy intervened and blocked them. The Navy had come to believe that it was not in the interests of the United States to permit a foreign-controlled corporation to control all its external communications by wireless and, in October 1919, they were able to bring about the formation of the Radio Corporation of America (RCA).

RCA was formed through the co-operation of a number of American companies that had interests in wireless, the principal ones being General Electric, AT&T, Westinghouse, and Western Electric. All these companies passed to RCA rights under all their wireless patents, including sub-licensing rights. General Electric was the principal shareholder in the new company and entered into an agreement with it, running until January 1, 1945, which transferred to RCA an exclusive world-wide license, except for Canada, for all inventions and patents for radio purposes which General Electric then owned or might acquire during the term of the agreement.

In the meantime, negotiations were underway between General Electric and MWT regarding the takeover of American Marconi Company and its incorporation into RCA. MWT's position was primarily aimed at protecting its interests in the British Empire and certain concerns in South America. Negotiators for General Electric, on behalf of the as then unregistered RCA, were attempting to widen as much as possible their control to give them a world-wide marketing base. Canada was a special problem in these negotiations, as neither side wished to allow the other control in an area that was at the same time part of the British Empire and in such close proximity to the United States. At one point during these discussions, it was proposed that RCA take an equity position in CMC, but this was not pursued.

On November 21st, 1919, an agreement was finally concluded between RCA and MWT, running until January 1st, 1945. (RCA had, in the meantime,

> *Perhaps I should intervene at this point to make quite clear that in those days one of the most serious problems that design engineers and people in the factory had to contend with, was the lack of materials. When I refer to lack of materials I am of course thinking of the fact that we had as insulators really only four things: One was fiberboard which is very hydroscopic and hence unreliable as an insulator and it develops all sorts of warps and twists which is very bad from a mechanical point of view. The second was hard rubber which flows under pressure and hence, if you for example tightened up some terminal tonight, tomorrow it will be quite loose again. The third was slate, which was certainly firm, durable and that sort of thing but also very heavy and somewhat awkward to handle. And then finally we had various forms of porcelain insulators which were used in aerial work and the like where very high voltages were concerned. Nowadays we can turn to a wide variety of materials each of which has been designed to serve a specific purpose and which in general whether they be insulators or otherwise are stable under all sorts of temperature, humidity and other conditions. So I plead with you not to think that when the so called wireless designer of 50 or more years ago worked out some of their relatively crude devices that you may see in museums and the like, that he was necessarily off base. On the contrary, he had fewer tools to work with and the results of his work were very severely limited by this factor.*
>
> **Unpublished recollections of S.M. Finlayson, CMC President from 1951 to 1964.**

become duly registered in the state of Delaware.) British Marconi was to retain exclusive rights in countries of the British Empire, except that for Canada a special agreement was to be drawn up under which Canadian radio stations would be restricted to Canadian business, and giving RCA an opportunity to acquire a substantial interest in CMC should it so desire. RCA was to hold similar rights for most countries in the Western hemisphere, and the rest of the world was to be treated on a competitive basis. At the same time, the parties to the agreement expressed their intention that RCA would execute an agreement with CMC concerning traffic arrangements and patents, but no evidence has been found that this agreement was ever completed.

In the period immediately following WW I, there was a marked increase in Marine activities in Canada. Many new ships - in some cases whole new fleets - were built; for example, the Canadian Government Merchant Fleet was established with approximately 70 vessels, all of which were fitted by CMC with equipment of its own manufacture.

At the end of 1919, the Company purchased the building at 11 St. Sacrement Street in Montreal, and all general administrative and clerical activities were transferred to that location.

During the early part of April 1920, new Marconi wireless telephone equipment became available, and Mr Barrow was assigned to give the first wireless telephone demonstration in Toronto, from the Canadian Marconi Company's office in that city, to the residence of Senator Nicholls. Guests present at the latter location included members of the Cabinet and the House of Commons. On May 20, Mr Barrow went to Winnipeg to conduct two-way wireless telephone demonstrations for the first time in that city, and in Portage-la-Prairie, some 66 miles to the west. The tests were held between May 25 and June 25, 1920. Mr Reginald Scantlebury and Mr Barrow set up two stations, one in Kelvin High School, Winnipeg, and the other three miles away in the supply department of Manitoba Government Telephones. Several Provincial and Civic officials and representatives of the Press were among the onlookers.

A second Winnipeg demonstration was conducted between the Technical High School and

Fig. 36. Original XWA station, Montreal, 1921

Portage-la-Prairie. Mayor Metcalfe of Portage, President Wood of the Portage Board of Trade and the Editor of the Daily Graphic spoke with Mr G.A. Grierson, Minister of Public Works, and other officials in Winnipeg. Tests were also made for the Winnipeg Power Company with Chief Engineer Caton attending, the stations being set up in the power plant and at Pointe-du-Bois, Manitoba. Mr George A. Watson, Commissioner of Manitoba Government Telephones, was a constant observer throughout all the tests. Further demonstrations of the "Wireless Telephone" were also carried out in Toronto and Montreal.

An interesting point to notice with regard to these tests is that they were for the purpose of promoting the use of wireless telephony for point-to-point communication rather than for broadcasting entertainment programs to the general public. Music columnists of the Manitoba Free Press and the Winnipeg Tribune do not appear to have attended the tests. Nearly two years were to pass before the two newspapers installed broadcasting equipment and went on the air. The CMC, however, was still conducting tests with its experimental station, XWA (Figure 36).

On September 10th, 1920, an important agreement was concluded between MWT and Canadian General Electric Company Limited, which included the following provisions:
1. The holdings of CGE and MWT in CMC were to be equalized at 800,000 shares each.
2. Until December 31st, 1944, CGE were to be entitled to name a majority of the Directors of CMC.
3. CGE would grant to CMC an exclusive licence in Canada until December 31, 1944, covering all inventions and patent rights of CGE in the field of wireless telegraphy or telephony.
4. This licence excluded CGE from the use of wireless devices, but it reserved the right to manufacture them. CMC had the right to use and also to manufacture, but it could not pass to third parties manufacturing rights under CGE patents. (It seems quite clear that the Alexanderson alternator was the focal point of this agreement.)
5. CMC was to issue to CGE 1,000,000 fully paid common shares of CMC at a par value of $2.50 each.
6. CGE was to pay CMC $500,000 in cash.
7. CGE was to transfer to MWT about 200,000 fully paid shares of CMC.

In order to implement this agreement, another agreement dated October 1st, 1920 was completed between CGE and CMC which provided that:

1. CMC should forthwith issue to CGE 800,000 shares of CMC of a par value of $2.50 each for patent rights.
2. A further 200,000 shares of CMC were suscribed for and allotted to CGE at par, such shares to be paid up in cash as CMC might from time to time demand.

World War I had given a great impetus to the development of thermionic valves (or vacuum tubes, as they were coming to be called in America). This development resulted, at least as early as 1920, in the experimental use of valve transmitters, with output powers of a few kilowatts. The introduction of the vacuum tube greatly improved marine communications and created a demand for the installation of vacuum tube equipment on coastal stations and ships.

At about this time, CMC was giving serious consideration to establishing across Canada a chain of commercial stations that could receive trans-oceanic messages directly, thereby eliminating the necessity of transferring such traffic to land lines at Loiusbourg. It was proposed that these stations would also pick up traffic for trans-oceanic transmission. Land was acquired at La Prairie, Quebec; Toronto, Ontario; and Winnipeg, Manitoba, and a station actually was erected at La Prairie which later carried out tests. However, due to unforeseen difficulties the plan was later abandoned.

In 1920 also, it was decided to consolidate all marine operations, and a Marine Department was established in Montreal, with full responsibility for all marine activities throughout the Company. Mr. W.J. Gray was transferred from Halifax to take over the newly created post of Marine Superintendent.

By 1921, interest in radio programs had developed to a point at which CMC felt the need for a monthly publication that would inform its readers of general progress in broadcasting and serve as a medium for radio "Hams" and others interested in wireless as a hobby to report their activities and exchange ideas. In June, 1921, Canadian Wireless Magazine, under editor Darby Coats, was issued as a four-page folder, price ten cents, "post free to subscribers". The August, 1921 issue was enlarged to eight pages, and by July, 1922 had grown to a 20-page periodical, listing some broadcasting stations in Canada and the United States and reporting items of interest to home-radio builders and listeners.

During October 22-24, 1921, the Canadian challenger Bluenose defeated the American defender Elsie in a series of races held off Halifax harbour. Special arrangements had been made between the

Fig. 37. Glace Bay, Nova Scotia, valve transmitter

Canadian Press and CMC, with the result that two special continuous-wave transmitters and receivers were installed, one on the official press boat, the Tyrian, and the other in the Halifax offices of Canadian Press. Communication by Morse code was kept up continuously throughout the races and enabled Canadian Press to bring the results to a wide audience in record time. At the end of each day's races, radio telephony transmissions were also made with great success.

Another important event which occurred in 1921 was the institution of a valve transmitter for continuous wave operation at Glace Bay Station, using 24 MT6 air-cooled transmitting valves (Figure 37), later expanded to 48. In addition, 48 rectifying valves were used; the spark equipment was retained as a stand-by. This system operated on a wavelength of 7600 meters.

In 1921, a new agreement was concluded with the Canadian Government covering the operation and maintenance of eight stations on the Great Lakes and 18 stations on the East Coast and the St. Lawrence River. This agreement was made for a period of five years and superseded all previous agreements for the operation of Coast Stations. It provided for an annual subsidy of $5,500 for each station, and CMC was still entitled to retain all Coastal Station tolls. The agreement also stipulated that CMC would charge only half the commercial rates for all Canadian Government messages.

During the latter part of the year 1919, ships almost halfway across the Atlantic were communicating directly with Barrington on the south coast of Nova Scotia and with WCC at Cape Cod, Massachusetts, thus eliminating the toll charge from

I spent quite a bit of time at Glace Bay and its associated receiving station at Louisbourg in the years from 1922 to about 1926 or '27, partly installing and rearranging equipment and partly at what was then called a shift engineer. That involved keeping the apparatus in working condition so that the operators would have full use of the circuits to Europe during the whole day. This apparatus was very different from today's. At the time I went to Glace Bay, the spark equipment had very recently been supplanted by a so-called valve transmitter which contained, at the beginning 24 and later, 48 valves each with its own individual grid leak condenser system. Any technical man will know what a very unstable device this was. Incidentally it was a self excited animal without benefit of any master oscillator crystal control of frequency and that sort of thing. It operated on 7,600 meters, roughly 40 kilocycles, which is a very different sort of thing to the frequencies used in today's long distance communications.

The then recently outmoded spark equipment was rather a fearsome and wonderful thing to see and especially to hear. On a still day, the spark itself could be read by ear five miles away, and that will give you some idea of the noise that existed in the condenser house where the gear was installed. They also at that time, prior to 1920, had a 20,000 volt wet cell battery which was used as a sort of a stand-by, charged part of the day and then discharged by telegraphing during the rest of the day. Anyone who loved batteries as little as I did could have a real nightmare contemplating the upkeep of some 10,000 individual cells.

Unpublished recollections of S.M. Finlayson, CMC President from 1951 to 1964.

local stations along the coast such as Cape Race, Sable Island, and Camperdown. During the years 1920 to 1922, these activities intensified to a considerable degree and the Company set up station VAS ("Voice of the Atlantic Seaboard") with a high-power continuous wave transmitter in Glace Bay and a receiving station in Loiusbourg, Cape Breton, which permitted it to offer shipping companies a duplex automatic high-speed service up to 1,500 miles off-shore. In 1921, Loiusbourg was also the site of a demonstration of the practicality of operating two centrally controlled wireless stations on one site, the so-called "radio central" concept.

In 1922, CMC equipped its first formal broadcasting studio, located in the Canada Cement Building on Phillips Square, Montreal, and the Canadian Government assigned it the call letters CFCF, which are still being used. Radio was in its infancy at that time, but a great deal of interest had been awakened, and many studio and remote programs were broadcast from that location. Later, the equipment for what is now Canada's senior French-Canadian station, CKAC, was assembled and tested in a corner of CFCF's new studio prior to its delivery to La Presse newspaper. Marconi radio engineer Leonard Spencer worked on the set in CFCF's quarters and became engineer-in-charge in the La Presse building.

Marine operations continued to expand during this year and the first direction-finding equipment which was developed by the Company's engineers, was installed on ships (MDFR-1, Figure 38). During this year, also, a service was established at Glace Bay Station to supply weather reports and news to ships.

At this time, Great Lakes ship spark transmitters and Coast Stations spark transmitters were changed from 600 to 450 meters, as the former interfered badly with broadcast reception. This change made it necessary to rebuild certain parts of the equipment.

An event occurred in 1923 that was to have an important bearing on the future activities of the Company for many years. This was the signing of agreements on March 7th with CGE and the Radio Valve Company of Canada Ltd. (RVC). The latter company had been incorporated in November 1922 for the purpose of acquiring, as a going concern, the business of CGE in connection with the manufacture and sale of thermionic valves, which had previously been carried on by CGE at its factory in Toronto.

The most important provisions of the agreement may be summarized as follows:

1. CMC and CGE would each suscribe in equal ratio for the Capital Stock of RVC. (The amount actually subscribed totalled $100,000 each.)
2. CGE would be entitled to nominate three of the five directors of RVC.
3. RVC would be entitled to receive the full benefit of all patent rights, present and future, belonging to or controlled by either CGE or CMC for a term up to December 31st, 1944.
4. RVC undertook not to supply or sell thermionic valves to any entity other than CMC but the latter agreed to ensure that CGE would be supplied with its requirements at a price equal to that charged to CMC plus a profit of 10%.
5. CMC agreed to purchase all its thermionic valves from RVC.

It appears that the agreement on the part of CGE to pay CMC 10% profit on all valves supplied to CGE was mainly due to the dominant patent position of CMC. Also, of course, CGE had at that time a large shareholding in CMC. In any event, this arrangement remained in effect until 1944 and provided a substantial source of revenue for the Company.

Another important agreement, known as the "Canadian Radio License Agreement" was also concluded in 1923, signed by CGE, MWT, the Canadian Westinghouse Company Ltd., The Bell Telephone Company of Canada Ltd., the Northern Electric Company Ltd, International Western Electric Company Inc., and Rogers Majestic. It provided for cross-licensing under the patents owned or controlled by the parties to the agreement and was drawn for a minimum

Fig. 38. Marconi Direction Finding Receiver model MDFR-1

period of 10 years, after which any one or more of the parties could withdraw by giving one year's notice in writing. However, with respect to CGE only, the agreement was to be terminated in any case at December 31st, 1944. Actually, the basic principles established by the agreement continued to be effective, with some minor changes, until at least 1956.

In the Marine Department, 1923 saw the beginning of the conversion from spark to the new 4KVA CW-ICW transmitters in all Canadian Government Coast Stations. These new sets were designed and developed by CMC, and the change-over program extended over a period of about two years. The first experimental station was established in Montreal, but the first production equipment was installed at the wireless station in Toronto. Both the Montreal and Toronto stations had 2kW plate input, all others having 4kW.

Broadcasting was continuing to gain popularity as new receivers for domestic use became available, but this activity had not yet developed to the point of producing revenue. In 1923 CMC produced its second domestic broadcast receiver, the Marconiphone Mk I (Figure 39). The set consisted of two boxes with sloped front panels, a tuner/regenerative detector and a two-stage audio amplifier. The list price was $50. In 1924 an electrically equivalent set was introduced, the Marconiphone Mk II in a larger and different style cabinet (Figure 40). The list price in 1924 was $90. The complete record of CMC domestic receiver production from 1921 to 1931 is shown in Table 2.

The year 1924 also saw the introduction of a series of receivers with tuned radio frequency (TRF) designs. The first of these was the Marconiphone Mk III, a sloped front receiver of substantial proportions (the panel was 14.5" x 22") whose presence would have been noted in anyone's parlour (Figure 41). It used five tubes, and its price in 1924 was also substantial at $300.

Fig. 39. Marconiphone Mk I receiving set

Fig. 40. Marconiphone Mk II receiving set (Robert Murray collection)

Fig. 41. Marconiphone Mk III receiving set (Robert Murray collection)

Fig. 42. Marconiphone Mk V, Queen Anne model of 1924

Table 2. Canadian Marconi Company Domestic Receivers, 1921-1931

MODEL	FIRST PRODUCED MM	FIRST PRODUCED YYYY	# TUBES	LIST PRICE	PWR	CCT TYPE	BAND	STYLE	COMMENTS
"C"	U	1921	1/3	$105.00	DC	Regen		Table	ST-1 Tuner; VD-1 Det; AA-1 Amp
Mk I	U	1923	3	$50.00	DC	Regen		Table	
Mk II	U	1924	3	$95.00	DC	Regen		Table	
Mk III	U	1924	5	$300.00	DC	TRF		Table	
Mk IV	U	1924	4	?	U DC	TRF			Chassis only
Mk V	U	1924	7	$600.00	DC	TRF		Cabinet	Queen Anne
Mk VI	7	1925	4	$115.00	DC	TRF		Table	
Mk VII	7	1925	8	$250.00	DC	Super		Table	
Mk VIII	8	1926	5	$125.00	DC	TRF		Table	
Mk IX	8	1926	6	$115.00	DC	TRF			Console
Mk X	8	1927	7	$325.00	AC	TRF		Console	
11	8	1928	6	$160.00	DC	TRF		Console	
12	8	1928	7	$265.00	AC	TRF		Console	
13	9	1928	8	$475.00	AC	TRF		Console	"Thermionic"; Combo
14	8	1929	7	$160.00	DC	TRF		Console	"Triumph" Series; 6V
15	8	1929	9	$175.00	AC	TRF		Table	"Triumph" Series
16	8	1929	9	$268.00	AC	TRF		Console	"Triumph" Series
17	8	1929	9	$365.00	AC	TRF			"Triumph" Series; Combo
18	8	1929	9	$890.00	AC	TRF		Auto	"Deluxe Triumph"; Combo
19	7	1930	7	$150.00	DC	TRF			"Superior"; 2V
20	7	1930	8	$185.00	AC	TRF			"Junior"
21	7	1930	7	$225.00	AC	TRF			"Standard"
22	8	1930	9	$285.00	AC	TRF; AVC			"Senior"
23	8	1930	9	$385.00	AC	TRF		Console	"Senior" Combo
24	12	1930	8	$295.00	AC	TRF			"Junior" Combo
26A&B	7	1931	7	$89.50	AC	Supe		Mantel	
26C	7	1931	27	$119.50	AC	Super		Console	
26SW	8	1931	9	$164.50	AC	Super	S	Console	"International"
27SW	11	1931	11	$198.50	AC	Super; AVC	S	Console	"Deluxe International"
28	11	1931	4	$49.50	AC	TRF		Mantel	"Bantam"

Note: The entire production table until 1950 has been published by Roger Hart in the Ottawa Vintage Radio Club Newsletter, 1999, Volume 7, Numbers 1 to 3, and as a supplement to Radio Waves, the Publication of the Canadian Vintage Radio Society 1999, Volume 7.

Little information remains about the Marconiphone Mk IV, except that it was also a TRF set of 1924, and used four tubes. The Marconiphone Mk V (Figure 42), also of 1924, appears in its original advertising photo to be of a similar panel layout to the MK III. It is again a TRF housed in a stylish floor cabinet, and is described as the Queen Anne model. It used 7 tubes and cost $600 in 1924!

For many years prior to 1924, the subject of a communication system to serve the entire British Empire—the so-called Imperial Wireless Scheme—had been under discussion in England and between the British Government and Governments of other Empire countries. As early as 1913 a plan had been completed for the establishment of a network of stations in different parts of the Empire, but the outbreak of WWI brought everything to a halt. Following WWI there was a great deal of further discussion on this subject, but it was not until 1924 that a decision was reached to proceed, on the basis that the British Post Office would own and operate all wireless stations in Great Britain communicating with the Empire.

In the meantime, experimental results had convinced Marconi and his engineers that a new means of long-range communication was now feasible which would completely outmode the long-wave high-power stations that had originally been intended for the scheme. This revolutionary method was to achieve the required performance by means of short-wave "beam" stations, which would cost much less than the original design, use only a fraction of their power, and have a theoretical capacity of up to 400 words per minute.

The proposal to erect the new beam stations was put before the Governments of the various Empire countries concerned by MWT and was unanimously accepted. As a result, CMC entered into an agreement with MWT which provided that the latter would construct the Canadian stations, CMC acting as their agents for this purpose, and that upon completion, CMC would purchase the stations from MWT on what amounted to a cost-plus basis. Sites were acquired in Yamachiche and Drummondville for the receiving and transmitting stations respectively, to be connected to a Central Telegraph Office to be established in the Marconi premises on St. Sacrement Street in Montreal (Figure 43). The final cost of these stations was in excess of $1,000,000.

During 1924, the first revenue was received from the operation of broadcasting station CFCF.

The construction of the new Beam Stations went forward rapidly in 1925, and negotiations with the British Post Office towards an agreement for handling traffic between Canada and England were started.

During the period 1923-1925, several new CW/ICW valve transmitters (type 100 W3, W4, and W5 sets) with power outputs of 50 to 500 Watts were installed in Canadian coastal stations. One Alexander McLean, having a bad hair day, is shown operating the Belle Isle station in the winter of 1925 (Figure 44).

Under date of June 12th, 1925, an amendment was issued to the Company's Act of Incorporation, which provided for the following changes:

1. The name of the Company was changed to Canadian Marconi Company.
2. The par value of the Capital Stock was reduced from $2.50 to $1 per share, and the authorized Capital increased to $7,500,000.
3. The Company was permitted to conduct a business of telephonic communication by means of wireless telephony and to transmit telegraph and telephone messages for the public and collect tolls therefor.

Fig. 43. The opening of the Transoceanic Beam Service from Montreal, August 1926.

Fig. 44. Alexander McLean operating Belle Isle Radio, winter 1925.

Further progress was made in the production and sale of broadcast receivers and accessories. The Marconiphone VI broadcast receiver was introduced in July of 1925 (Figure 45). It was another TRF set in a large (28" wide} table cabinet. It consisted of four tubes and cost $115. Also in July of 1925 came the Marconiphone VII, CMC's first superheterodyne (Figure 46). It used 8 tubes and cost $250. Sales must have been disappointing because CMC did not try another superhet until model 26 in 1931. Broadcasting activities also increased, and CFCF earned a total of approximately $12,000.

Fig. 45. Marconiphone Mk VI receiving set

The new beam stations were officially opened on October 25, 1926, when the circuit between Canada and England was put into operation. This inaugurated a new era of wireless communication, which later led to the extensive "Via Marconi" Empire network. Also, in the same year, an agreement with the British Post Office, covering the handling of traffic on the circuit between Canada and England, was concluded.

During 1926, negotiations were entered into with both the Canadian and Newfoundland Governments with a view to obtaining a revision of the subsidy for the operation and maintenance of Coast Stations owned by these Governments. Also in 1926, a new 500 Watt CW-ICW long-wave transmitter was pro-

Fig. 46. Marconiphone Mk VII receiving set

duced and installed on ocean-going vessels.

With the opening of the beam service, the Company's high-power station at Glace Bay, previously used for communication with England, was closed down and the cost of the necessary transfer of operations was written off during the year.

In August of 1926, Marconiphone receivers models VIII and IX were added (Figures 47 and 48). Model VIII used 5 tubes and sold for $125, while model IX used 6 tubes and sold for $115. Both used a TRF design. Looking at these models today, it is difficult to see why the model VIII should have been the more expensive of the two.

In 1927, 2,250,000 shares of CMC stock were sold to a holding company in which Robert Brand and Sir Robert Kindersley, on behalf of Lazard Brothers Ltd., were the major force, with the Radio Corporation of America a non-voting partner. The new company was incorporated on June 25, 1927 under the name "Canmar Investment Company Limited" with an authorized Capital of 5,000 Class "A" shares at a par value of $1 each, and 18,500 Class "B" shares at no par value. Voting rights were held exclusively by the Class "A" shares. MWT received 1,000 Class "A" shares as part payment for the CMC shares sold to Lazard Bros, which of course gave them control of Canmar and indirectly of CMC. This situation existed until 1940 when the interests of Lazard Bros and MWT were acquired by Cable and Wireless Ltd.

The main objectives of this agreement appear to have been to ensure that ownership and control of CMC would remain in British hands, and to provide that RCA and its connections would be the normal routing for all radio traffic received by CMC from stations of the British Empire radio chain or other places, destined to the United States and other countries of the American Continent south of Canada. RCA and Lazard wanted a CMC alliance with the Canadian Pacific Railway (CPR) in order to forestall the possibility of CPR (Canada's leading telegraph company) joining with American telegraph companies in a competing wireless venture. CPR did, in fact, acquire a small interest in CMC, but not before CMC had switched its receiving and distribution contract to the Canadian National Railway (CNR), (effective November 1, 1928).

The relationship between the par value of CMC shares and the trading price during this period is interesting. In 1927, when CMC stock with a par value of $1 was selling on the New York Curb Exchange at around $3, the company president, J.W. Flavelle

Fig. 47. Marconiphone Mk VIII receiving set (Robert Murray collection)

Fig. 48. Marconiphone Mk IX receiving set

issued a statement saying that the price was unwarranted. Still, by November 1928, the price had risen to over $28.50, giving the Company a market value of $129,789,000, despite the fact that it had assets of only $5,000,000 and earnings of $.01 per share! Deeply concerned by this situation, Flavelle gave an interview with Floyd Chalmers of the Financial Post in which he repeated his earlier statement that the stock had been overpriced even at $3.00. In two days, it fell from $28.00 to $8.00, then rallied to $9.00 to $10.00, still 900 to 1000 times earnings.

On July 1st, 1927, an historic event took place in which the Company's beam station at Drummondville and station CFCF in Montreal played an important part. This was the great Jubilee Broadcast from Parliament Hill in Ottawa, which was arranged to commemorate Canada's Diamond Jubilee. Through the co-operation of the telephone and telegraph companies and broadcast stations throughout the country, the broadcast was heard across Canada, and the transmission from Drummondville was heard in many other countries. H.M. Short, then Managing Director of CMC, was a member of the Diamond Jubilee Broadcast Committee.

Arising out of the Canadian Radio License

Agreement of 1923, a new agreement was drawn up dated January 19, 1927 that covered the incorporation of Canadian Radio Patents Ltd., established for the purpose of pooling the patents of the shareholding companies and issuing licenses to other Canadian manufacturers on a royalty basis. Under this agreement, CGE, CMC, Canadian Westinghouse Company, Ltd., and Northern Electric Company Ltd., each received 6,250 shares of the new company, and Standard Radio Manufacturing Corporation Ltd., received 2,500. The par value of the shares was $1 each.

In the Marine Department, short-wave transmitters were designed and installed for point-to-point communications. By that time, government regulations made it compulsory for certain types of vessels to carry automatic alarm devices, enabling them to dispense with one of their operators, the 'watcher'. The auto-alarm operated a bell when a signal of three 3-second dashes, spaced one second apart, was received. Twelve such dashes and spaces were always sent out before an SOS signal. CMC manufactured these devices, and benefited greatly from sales to meet the new requirement.

The volume of business in broadcast receivers and valves continued to expand and the Company produced its first AC receiver, the Mark X in August, 1927 (Figure 49). This was a 7-tube TRF set costing $325. By this time, broadcasting was becoming established as an advertising medium, although CFCF was still operated as a branch of the Sales Department. In June, 1927, the station was moved from the Canada Cement Building to the Mount Royal Hotel, where it remained for several years.

In 1928 an Empire Government Conference was held in England and resulted in the formation of a merger company to combine the respective interests of Cable and Wireless and MWT in Great Britain. This new company was called "Cable and Wireless Ltd." and had as its object the acquisition of the stocks, shares, debentures and other obligations of the various companies concerned, including MWT. In addition, a separate communications company was formed under the name "Imperial and International Communications Ltd.". This company acquired the physical assets of the same group of companies relating to communications, in exchange for shares, thus segregating in one company the communications aspect of the undertaking. The inclinations of Joseph Flavelle, President of CMC, at that point

> *The factory was concerned with basically two things. The supply of equipment for our marine services both on ships and on coast stations, and allied to it a certain amount of broadcast equipment which was, in part at least, manufactured in our factory at William Street. The other major effort was concerned with production of broadcast receivers, which at that time tended to be produced only from about the beginning of July until around Christmas time. Then there was a complete lull and absolute cessation of production from about the beginning of the year through until early Summer. This period, although an idle one from the factory point of view was a very busy one from the design point of view. It was during the early months of the year that the sales people provided the design engineers and the factory with their ideas as to what the market would require and then there were the usual processes of technical design, the preparation of drawings, the preparation of models, certain kinds of field testing and finally an authorization for production. In those days, the quantities manufactured were relatively small and as I recall our output was limited to 50 or 100 sets a day at the peak.*
>
> **Unpublished recollections of S.M. Finlayson, CMC President from 1951 to 1964.**

Fig. 49. Marconiphone Mk X receiving set

seems to have been to sell CMC to I&IC, thus retaining British control. This move, however, appears to have been blocked by David Sarnoff and Owen Young of RCA, who felt that a new manager should be found for the company and that it would then be capable of a greater degree of independence, and by Canadian Prime Minister R.B. Bennett, who was diametrically opposed to greater British control of Canadian industry. As a result, Flavelle offered to resign, but was convinced to stay until 1934, when he handed over the post of President to Perry, the same year in which CMC was, in fact, finally acquired by I&IC. Two years later, Flavelle resigned his post as Chairman of the Board and retired. The Beam Services were extended in 1928 by the opening of circuits for communication with New York and Australia. Also during this year CMC's manufacturing business continued to increase and the first broadcast receiver components were manufactured.

Broadcasting activity continued to expand as its value as an advertising medium became more widely recognized and the gross revenue from CFCF increased in 1928. Broadcasting, however, was having a difficult time getting established in Canada, as the economics of the business were at best difficult. As historian Michael Bliss puts it, "The trouble with Canadian radio was American radio. Powerful American stations had always covered Canada more thoroughly than home-grown transmitters. With the creation of national radio networks by the end of the 1920's, the Americans began broadcasting hugely popular entertainments that no Canadian producer could possibly match".

1928 also saw the first of many transatlantic broadcasts, when a Thanksgiving Service for the recovery of King George V was broadcast by the BBC from Westminster Abbey in London, England. This transmission was picked up by the CMC station at Yamachiche and sent all over Canada via the CNR network. The Thanksgiving Service was followed on November 11 of that year by the Armistice Service from Whitehall, and two months later by a live broadcast of King George V speaking at the opening of the Navy Parley in London.

The year 1928 also saw the introduction of three new broadcast receivers, models 11, 12 and 13. Model 11 was a 7-tube TRF set, battery operated, that cost $160 in a console cabinet. Models 12 and 13 were 7 and 8 tube sets respectively, both TRFs, costing $265 and $475 in console cabinets. Both were AC operated.

In the Broadcast Receiver Division, sales continued to increase for the first nine months of 1929 and

> *I well remember that our whole shipping department would be completely plugged by say 30 or 40 receivers at one time. It must also be appreciated that although these receivers were in both the mantle or table form, and in the self-supporting cabinet form, they were all very much bigger than what was to be the norm in later years, at least until we got into the production of Hi-Fi units. The reason for this was that the market seemed to demand a larger unit than was the case in the 1930's, and also because structurally it was not possible to reduce the size very much. One of the major elements was that in the very late twenties we began to include loudspeakers and these in themselves were quite large, probably 18" x 18". The components available at those days were much larger than we are accustomed to in recent times, and we had a number of deficiencies I suppose you would call it in the methods used in assembly. Consequently space was wasted as compared with later years.*
>
> *Our receivers of the self-supporting cabinet form were crated in what were then called wirebound boxes, which in turn were rather cumbersome things. Certainly a lot more awkward and difficult to deal with than the modern carton, with its very appropriate packing elements. In many cases we had difficulty in storing in the limited space in the factory the components and in particular the cabinets that were needed from day to day. Consequently we had to work on a somewhat hand to mouth basis, depending upon suppliers to deliver perhaps twice a day or maybe even three times a day in order that our incoming and outgoing shipping, boxing and unloading facilities would not be completely swamped.*
>
> **Unpublished recollections of S.M. Finlayson, CMC President from 1951 to 1964.**

the first combination radio and record player was produced (Figure 50). The dynamic speaker was also introduced in 1929. Towards the end of the year, however, the depressed state of the economy resulted in a sharp drop in demand for domestic receivers, and inventories grew rapidly.

There was a further expansion of broadcasting activities in 1929, and the gross revenue increased to almost three times that of 1928. The net profit on the operation of CFCF also increased.

From the Company's point of view, the most important event in 1930 was undoubtedly the move of the production facilities on May 1 to a new factory that had been erected in the previous Fall and Winter in the Town of Mount Royal, a suburb now, in 2001, close to the centre of Montreal, at a cost (including land) of approximately $275,000. This new plant represented the latest in modern manufacturing efficiency, and the public was invited to the official opening, which took place during the week of September 8th to 13th. The original land consisted of three acres between Canora Road and the Montreal city line.

The rear section of that area was at the time owned by the CNR, who had reserved it for a line leading to a team track to be erected at Rockland Avenue. The original parcel was purchased for 15 cents per square foot; a subsequent parcel was purchased for a mere 5 cents per square foot when the Town of Mount Royal sold it to recover unpaid taxes.

The move to the new facilities marked the beginning of a new era in CMC's development, as it provided much improved working conditions as well as ample room for future expansion. The Head Office of the Company, however, remained in the Marconi Building at 11 St. Sacrement St. in Montreal.

In the Marine Department, a new agreement was entered into with the Canadian Government, covering the operation and maintenance of Canadian Government Coast Stations for a period of one year on the basis of cost plus 10%. Also, during 1930, short-wave radiotelephone service was established between Glace Bay Station and ships at sea. Revenues from station CFCF continued to increase.

An interesting event in 1930 was the unveiling of a monument to General Wolfe, the victor in the 1759 battle to take Canada from the French, at his birthplace in England. The ceremony was broadcast on a national network in Canada, in which CFCF was the key station, and the Marconi facilities at Yamachiche were used to send the broadcast across the Atlantic.

In 1930 also, CFCF became an affiliated station of the National Broadcasting Company (NBC) and so had available to it most of the programs originated by NBC on its Red and Blue networks. It is interesting to note that up to the time of this affiliation, listeners in the Montreal areas spent much of their time tuning in American stations. However, after the advent of NBC programming on CFCF, they stayed with the local station and not only heard the programs of foreign origin but were content to listen to the majority of the programs which were of local origin.

The total sales of manufactured products increased during the year but higher distribution costs reduced the Net Profit. There was also a loss shown on the operation of the Beam Stations, but operations for the year resulted in a small net profit. On the domestic side, fierce competition forced price reductions, with a resultant decrease in margin.

This was the depression period and business was carried on under very difficult conditions. Sales revenue dropped to less than 50% of the 1930 volume and despite every effort to reduce costs, the manufacturing and selling operations showed losses in each of the three years. Operations in other departments were similarly affected, with the result that

Fig. 50. Marconiphone Mk 13 radio-gramophone combination

Fig. 51. Marconi model 26 broadcast receiver (Robert Murray collection)

the Company showed losses in 1931, 1932 and 1933.

Five new broadcast receivers were introduced in 1929. These were models 14 (a 7-tube battery console for $160), 15 (a 9-tube table set using AC, for $175), 16 (a 9-tube console using AC, for $268), 17 (a 9-tube radio-record player combination for $365), and 18 (a 9-tube deluxe combination for a whopping $890!). All of these sets were TRF. Evidently the year's models were designed and released (in August) before the impact of the depression was felt.

Six new models were released in 1930, models 19 to 24. All were again TRFs. CMC did not return to manufacturing superhets until the following year, 1931. The 1930 model prices ranged from $150 to $385, a slight decline from the 1929 prices. CMC's first mantel (or "Depression") set did not appear until 1931—the model 26 (Figures 51 and 52). To describe this set as small is a euphemism, since it stands over 18" tall. Its price in 1931 was $89.50, and it was a 7 tube superhet. CMC had not charged a lower price than this (not adjusting for inflation) for a receiver since the Mk I in 1923. Also in 1930 there was a model 28 TRF 4 tube "Bantam" mantel set for $49.50.

In 1931, CFCF, now located in the King's Hall Building at 1231 St. Catherine St. West in Montreal, was the moving spirit in the creation of Canada's first radio network. This comprised a group of stations in London, Ont., Hamilton, Toronto, Ottawa, Montreal, Moncton, Halifax, and Fredericton. The CNR, which owned stations in Ottawa and Moncton, contributed its telegraph lines and the stations in turn made programs available to the network.

Epilogue

Canadian Marconi survived the depression of the 1930's, although its revenues fell like those of many other companies. The advent of WWII led CMC into the military communications sector, where it made a major contribution to the war effort with equipment such as the famous No. 19 "tank" transceiver. After the war, product lines were diversified to include commercial radio equipment, military electronics, and avionics—electronics apparatus for aircraft. By the early 1960's CMC was deemed to be in conflict of interest due to the fact that it both owned a broadcasting station (CFCF) and manufactured domestic radio and television sets. Consequently, the

Fig. 52. Broadcast receiver assembly line, 1931, showing mostly model 26 mantel sets.

Company sold its broadcasting interests and ceased the manufacture of domestic equipment. The 1950's brought the development of new airborne systems, notably the Doppler Navigation System, sophisticated military communications equipment, and high-reliability electronic components. During all of that period, GEC in the United Kingdom held a controlling interest in the outstanding shares, and when part of that empire was sold to British Aerospace, the Canadian Marconi name disappeared, the Company being re-named British Aerospace Systems (Canada) Inc. (BAeS(C)I). By April, 2001, ownership had again changed hands, with the sale of BAeS(C)I to Onex Corporation. As of April 10, 2001, BAeS(C)I has been re-named CMC Electronics, Inc.

Photo credits

All figures are from CMC original photos, donated by Roger Hart to the Canada Science and Technology Museum, unless otherwise noted.

BIBLIOGRAPHY
(Some recent publications on the subject of Marconi)

Babaian, S.A. *Radio communication in Canada: A historical and technological survey.* Ottawa, ON: National Museum of Science and Technology, 1992.

Baker W.J. *A history of the Marconi Company.* London: Methuen, 1970.

Bartram, G. Marconi v. British Radio Telegraph and Telephone Company: the patent case that changed the world. *AWA Review,* vol. 13, pp. 39-80, 2000.

Bird, R. (Ed.). *Documents of Canadian Broadcasting.* Ottawa, ON: Carleton University Press, 1988.

Bliss, M. *Northern enterprise.* Toronto, ON: McClelland & Stewart, 1987.

Bussey, G. *Marconi's Atlantic Leap.* Coventry, England: Marconi Communications, 2000.

Lee, B. Marconi's transatlantic triumph. *AWA Review,* vol. 13, pp. 81-97, 2000.

Dubreuil, S. *Come quick, danger: A history of marine radio in Canada.* Ottawa ON: Fisheries and Oceans Canada, 1998.

MacLeod, M.K. *Whisper in the air: Marconi, the Canada Years,* 1902-1946. Hantsport, Nova Scotia, Canada: Lancelot Press Limited, 1992.

Masini, G. *Marconi.* New York, NY: Marsilio Publishers, (English edition), 1995.

Weightman, G. *Signor Marconi's Magic Box.* Cambridge, MA: Da Capo Press, 2003.

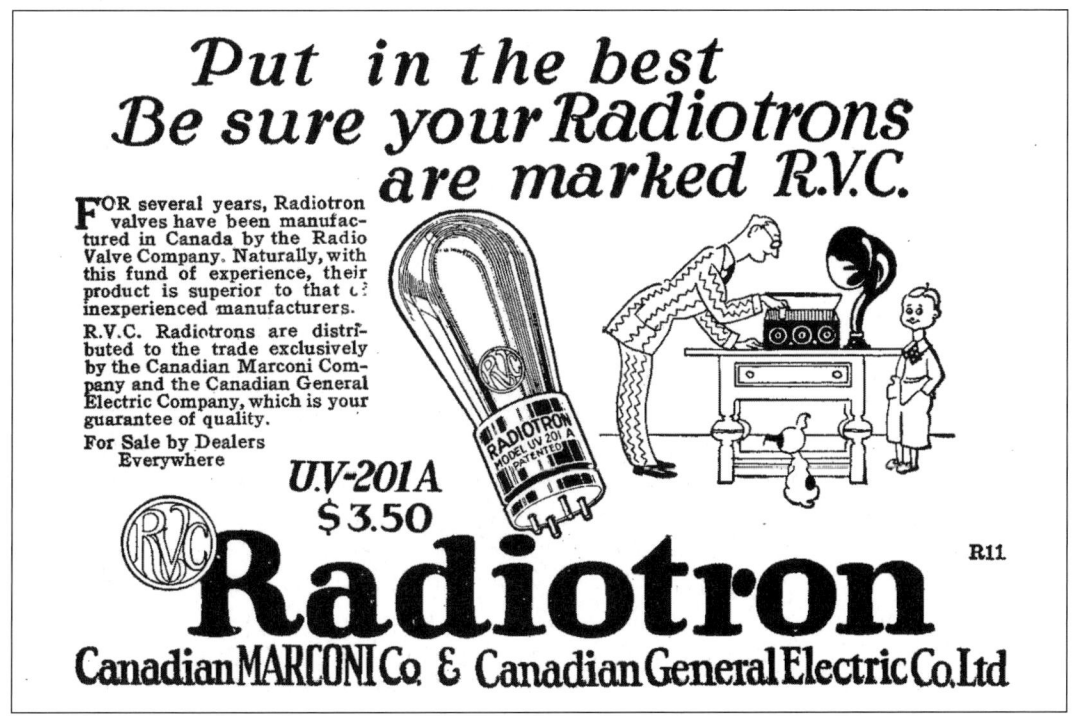

The Marconi Long Range Receiver

An Instrument of Proved Ability and Consistent Performance

Creative resources, unequalled experience, and manufacturing advantages without parallel, assure each possessor of this well-known model the best that Radio offers.

The Canadidan Forestry Association Exhibit Car, now on tour in the Maritime Provinces, was recently fitted with standard Marconi Equipment and reports as follows:—

"Concerts heard from Schenectady; New York; Davenport, Iowa; Omaha, Neb.; Detroit, Mich.; Chicago; Pittsburgh, etc.

"Music and voice reproduction were perfect on the Magnavox Horn and strong enough to be heard within a radius of 100 yards."

This car is fitted with a collapsible aerial averaging 5 ft. 6 ins. from the roof, which is of metal construction.

A Montreal user writes:—
"———'PWX', Cuba, is my farthest station and was heard through a Magnavox by four people."

Many other clients in different parts of the country report "PWX," Cuba, including one gentleman located in Cumberland, British Columbia, who receives this and other long distance stations on an indoor antenna.

We have prepared an interesting booklet containing a few selected reports from different parts of the country that make mighty interesting reading—your copy awaits you.

The Marconi Wireless Telegraph Company of Canada, Limited

Head Office and Factory

MONTREAL

The Development of Radio in Canada to 1928

A.F. Fraser, AMEIC
Chief Engineer, Radio Branch, Department of Marine, Canada

Canada, as befits a new and great country, has always been prompt to utilize new engineering and scientific developments, and in no field has this been more marked than in the way in which our Government and our people have made use of the art of wireless, or, as it later became termed, radio communication. And without distracting in the slightest degree from the transcending work of Mr. Marconi, in first conceiving the idea of using Hertzian waves for the purpose of communication, and then in developing the apparatus to attain this end, the importance of the contributions to the radio art of that brilliant Canadian engineer, Mr. Reginald A. Fessenden, born at Milton, Ontario, (born in Knowlton, Quebec, see Page 1. Ed.) are becoming more appreciated as the years go by, and it is a fitting tribute to Mr. Fessenden that within the last few months he has been chosen by the American Museum of Safety as the recipient of the Scientific American Medal for his contributions to the "Safety of Life at Sea".

Canada's first adventure into radio communication took place 29 years ago when in 1901, the late Mr. D.H. Keeley, a member of this Institution and Superintendent of the Government Telegraphs entered into a contract with Mr. Marconi to establish two stations in the Belle Isle Straits, one at Chateau Bay at the terminus of the landline on the mainland, and the other on Belle Isle, 30 miles distant, to replace the submarine cable which was continually going out of commission on account of icebergs.

It is a far cry from the thirty mile station of 1901, to the 10,000 mile station of today, but it was a still farther cry from nothing at all to 30 miles; and the promptitude with which Mr. Keeley seized on this means of communication at such an early date to solve his difficulties, indicates that he, in common with so many of the founders of our Institute, possessed the true spirit of the pioneer. In fact not only did Mr. Keeley make the contract, but he personally contributed to the success of the experiment. In his report of November 1901 he naively states: "This Chateau Bay installation was in readiness on Sunday the 20th October, when signals from Chateau Bay were received at Belle Isle, but none were received at Chateau Bay. The trouble, on investigation, was attributed to defective coherers, and the company's agents proposed abandoning the plant till next year in the absence of a fresh supply. On the 22nd, however, I personally succeeded in establishing communication and was thereby enabled to avert the threatened postponement of operations; and on the 25th after considerable practice and careful directions, for all that the working of the system as it stands is erratic, I felt confident in leaving our operators in charge, with the explicit instruction as to future action."

While in this experiment wireless telegraphy demonstrated its utility to replace a cable, it was in an allied field that it made its greatest contribution to progress and to the safety of human life, in providing the first means of communicating over long distances with a ship at sea; and accustomed as we are today to a condition where a ship is never out of communication wherever she may be in the world, it is difficult to appreciate for instance, that only a few years prior to the invention of wireless the S.S. Borassia left Liverpool with 344 souls on board, sprang a leak in mid-Atlantic and foundered; 160 persons went down with the ship, 184 took to the lifeboats of whom only 10 were eventually saved. It is of interest to note that the Marconi Company's records show that the first commercial equipment on board a liner was installed in 1901 on the Beaver Liner "Lake Champlain" plying from Liverpool to Montreal.

Those of you who heard Mr. Marconi during his trans-Atlantic broadcast a few weeks ago, will recall that the first trans-Atlantic signals were received at St. John's Nfld. by means of an aerial supported by a kite on December 12, 1901. And Sir Wilfred Laurier, the Canadian Prime Minister then in power, quick to appreciate the possibilities of wireless, invited the young inventor to proceed to Canada, with the result that a contract was entered into for the establishment of a trans-Atlantic wireless service between Canada and Great Britain, the Canadian Government to pay a subsidy of $ 80,000 towards the Canadian Stations.

Glace Bay, N.S. was selected as the site and in 1902 the first messages were exchanged between the Earl of Minto, then Governor-General of Canada, and His Majesty King Edward VIII, and after a somewhat prolonged experimental period, the first commercial trans-Atlantic service — in fact the first long distance service in the world — was inaugurated between the station at Glace Bay and one in Clifden, Ireland, in October 1907, and it might here be added that the inauguration of this service contributed in part, if not in whole to the fact that you are today paying a toll of only 15 cents per word, the rate previously in effect.

Concurrently with the contract for the establishment of the first Glace Bay trans-Atlantic station, contracts were entered into for the building of a chain of 20 coast stations along the River and Gulf of St. Lawrence, to the Belle Isle Straits and from St. John and Halifax to Cape Race along the East Coast. These stations were of comparatively low power and were installed in 1903, 1904 and 1905, and it is a tribute to the quality of the apparatus that at some of these stations in the more isolated points, where they do not cause interference, the major part of the original plant is still in operation.

In 1907 work was started on a chain of 9 stations to provide communications along the B.C. coast, including the Queen Charlotte Islands, and in 1912 a third chain of 8 stations was established from Port Arthur to cover the Great Lakes, incidentally linking up with the east coast chain in Montreal. At all these stations constant watch was maintained, day and night, 365 days a year, or throughout the season of navigation, as the case might be, and from the time a ship was 250 miles from our shores she was in constant touch with one or other of the stations until she reached her final port.

The original stations all carried about the same plant: the prime mover was a 4 H.P. engine and the transmitter was of the spark type rated at 2 KW. At first coherers and shortly afterwards magnetic detectors (the latter due to the research work of Dr. Rutherford of McGill University) were used for reception and the working range of this combination was some 250 miles.

As the art progressed, higher powered and more efficient sets using the same type of transmitter of 5 KW rating replaced the original outfits and with the more sensitive crystal type of detector the range of the stations was materially improved. Nevertheless the general design remained fundamentally unchanged, except in minor details, until in 1918, when transmitters employing vacuum tubes began to be universally adopted. Prior to this, radio engineers had long recognized that damped Hertzian waves, such as are created by an electric discharge across a spark gap, were by no means ideal for wireless communication. Such transmissions occupy a broad band in the ether and their carrying powers are limited.

To Mr. Fessenden before mentioned, was, I think, due the first practical method of producing a continuous wave, when in 1908 he built a high-speed alternator turning up to 20,000 revolutions per minute, which gave a frequency of 100,000 cycles (3000 metres).

The frequency of an alternator is, as you are aware, a function of the speed and as there are physical limitations to the latter, it was not found possible to develop an alternator which would produce electric waves of the frequencies then used for ship to shore communication of the order of 100,000 cycles per second. For long-wave long range work, however, the alternator was a success and the General Electric Company developed machines capable of an output of 200 KW with frequencies of the order of 30,000 cycles (10,000 metres) which were extensively used for trans-Atlantic and other long-distance services.

It might not be out of place to remark here that it was whilst experimenting with these continuous waves that Mr. Fessenden contributed his principle of heterodyne reception to the radio art, a principle simple but eminently effective and one which is used for reception at all continuous wave radio-telegraph stations today.

Marconi endeavored to obtain continuous waves by means of a timed spark arrangement, while Poulsen made use of the negative (resistance) characteristics of an arc in hydrogen. A large trans-Atlantic station of the latter type was built in Canada at Newcastle prior to the war and two or three smaller sets were installed by the Marine Department in coast stations. However, these were all temporary expedients waiting for the development of the vacuum tube, and such of them as are still in existence, will, together with the alternators, ere long be relegated to the historical museum along with the spark sets and be replaced by the three-element vacuum tube. This tube is due to the work of Dr. Lee de Forest of New York, who in 1907, experimenting with the two-element vacuum tube detector invented by Dr. Fleming, conceived the idea of introducing a third element into the tube, which he called the grid, and thereby he made possible the present phenomenal development in radio, though I question if at that time Dr. de Forest had the slightest idea of the revolution he was about to inaugurate.

It may perhaps seem strange that this tube can be used both for reception and transmission. However, every one of you who owns a radio broadcast receiving set probably has a nightly demonstration of the latter, when the owner of an neighboring one-tube regenerative "blooper" proceeds to whistle through your concert. He is using his receiving tube as a transmitter and it is made audible to you through Mr. Fessenden's heterodyning effect. A big transmitting station is merely a repetition on a large scale of what goes on in the one-tube regenerative set. With a vacuum tube transmitter it is possible to produce a pure continuous wave and at the same time to control it when using high power. As compared with a damped wave a continuous wave travels a much greater distance for the same expenditure of energy, occupies a very much narrower band in the ether and when modulated with a microphone it enables us to transmit sounds at voice frequencies.

The large investment in Canada and throughout the world in ship to shore spark equipment mitigated against the early adoption of the tube type transmitter, but insofar as Canada is concerned, the advent of broadcasting rendered such a change urgently desirable in the interest of broadcast listeners and the Canadian Marconi Company, having, on our behalf, developed a tube transmitter suitable for coast station operation, the Government embarked on a program of replacing all spark sets in its stations, connecting with those in the more populated centres such as Toronto, Montreal, Vancouver, etc., with the result that today, except in one or two isolated points, all our stations are operating on continuous wave.

Canada's pioneering in the matter of trans-Atlantic communication did not end with Glace Bay, and it is interesting to note that the first commercial long-distance beam short-wave radio service was that established by the Canadian Marconi Company between their station at Drummondville, near Montreal, and a station at Bodmin, England, which was put in commission on October 25th, 1926. Today hundreds of long-distance short-wave stations are in operation in practically every country in the world, and, as a consequence, one of the biggest problems with which those charged with the administration of radio are confronted is to fit them all in the radio spectrum. A station in Montreal interfering with a station in London which communicates with Montreal, is a much more complicated matter.

Prior to the advent of the beam in 1926, long distance communications were usually carried on long waves of between 8000 to 20,000 meters, with frequencies of 37,500 to 15,000, but since that date there has been a complete revolution in this phase of the art, and while today there are a few long-wave stations left, all new development is along short-wave lines.

Canada has two international short-wave circuits, the one to England above mentioned, and one to Australia opened by the Canadian Marconi Company in June, 1928. The frequencies used by the British circuit are 18,180 kc. day and 9,330 kc. night, and on the Australian circuit the same frequency is used both day and night, but at night signals are shot around the world the other way. In addition there is an experimental short-wave voice telephone circuit between Montreal and London which we look forward to seeing in commercial operation in the not too distant future.

Those of you who listened in on the speeches given at the inaugural session of the Disarmament Conference had an opportunity of testing the quality of this particular short-wave circuit, over which the signals were brought to Montreal, where they were put on the telephone wires and distributed to the Canadian broadcasting stations from coast to coast.

Telephone communication with Australia has also proved entirely practicable, but it is doubtful if there is sufficient commercial demand for such a circuit to warrant placing it in operation in the immediate future.

The Canadian Marconi Company's beam transmitting station in Canada is situated at Drummondville, thirty miles east of Montreal, and the receiving station at Yamachiche, twenty-five miles north of Drummondville. These stations are linked up by land line to the central office of the company at Montreal, from which the transmitter is automatically operated. The moment the operator in Montreal presses his key or feeds his message tape into a high-speed telegraph instrument, the signals he is sending are instantaneously recorded at the distant terminal office of the circuit, whether it be 3000 miles away, in London or the longer distance to Melbourne.

Incoming signals from the corresponding stations are received at Yamachiche and after being heterodyned to a lower frequency, amplified and filtered are conducted by landlines, consisting of open-wire lines and cables, to the office in Montreal where they are automatically recorded and are transcribed for delivery to the addressee.

By employing circuits analogous to those extensively used in carrier current telegraphy for superimposing several communication channels on one physical wire circuit each beam aerial can be efficiently utilized for the simultaneous transmission of telephonic or telegraphic messages without there being any mutual interference between these services. Facsimile transmission can also be employed with the beam system over practically any distance.

It may be interesting to note here that the carrier current systems of telegraphy now being extensively installed by both the Canadian Pacific and Canadian National Telegraphs on their long heavily loaded landlines for greatly increasing their traffic handling capacity employ apparatus made possible by the use of thermionic tubes, first developed by the radio industry. With regard to the relative advantages of the beam as compared with the older systems of long wave radio communication, the following may be cited:

1st: Less capital expenditure required.

2nd: Less electrical power required to operate the transmitters.

3rd: Greater speed of transmission possible. At present this is limited only by the mechanical limitations of the keying and recording instruments.

4th: Due to the restrictions of radiation to a narrow beam, to the screening effect of the reflector at the receiving station and to the large number of wave-bands available, a greater number of services can be carried on.

5th: Due to the screening effect of the receiving reflector the signal-to-interference ratio is increased and consequently the traffic capacity is increased because the possible sources of interference are reduced in proportion to the narrowness of the arc of reception.

The competition of the beam stations has introduced a new factor in international communication and as a consequence a merger has been formed in England to unite the British cable and radio telegraph systems into one.

One advantage enjoyed by the beam system over the cables is the fact that two beam stations capable of communicating half way round the world may be built for approximately half a million dollars whereas the cost of laying a permalloy loaded cable capable of working at similar speed is many times as great and increases in proportion to the distance separating the terminal stations.

In this regard it is interesting to note that a cable of this type laid between New York and the Azores, a distance of 2328 miles, is reported to cost $4,000,000.

Conclusion:

Even in this age of wonders the rapid progress made by radio stands out as phenomenal, and if we pause to consider its future possibilities, there is no doubt that, intelligently employed, it may become one of the most potent factors in bringing about mutual understanding among all peoples and the promotion of an irresistible sentiment in favour of universal peace.

~ 3 ~
The Telephone Industry

Roy Thompson, son of a Toronto barber, was a not-very-successful radio salesman, who began a North Bay radio station in 1931 with $500 worth of used Northern Electric equipment. He believed that owning a radio station would help him sell radios, and he turned out to be very effective in selling advertising. Two years later, at age forty, he stumbled into ownership of newspapers, again primarily for the chance to sell advertising. In 1984 International Thompson was worth over $10 billion, controlled over 200 newspapers in Canada, the U.S. and the United Kingdom, and made Ken Thompson, Roy's son, Canada's wealthiest man. (source: S. Goldenberg, The Thomson Empire. New York NY: Beaufort Books, 1984.)

The telephone industry is clearly older than the Marconi Company, but did not get directly involved in wireless telephony in a serious way until the advent of broadcasting. In the formation of the radio group in the U.S. in 1921, A.T.&T. was allocated the role of the manufacture of broadcast transmitters through its subsidiary Western Electric. The vast potential growth of broadcasting would have been much more apparent in 1923 when the Canadian patent pool was formed. A.T.&T. subsidiary Bell Canada secured a much larger role in the broadcast industry through its manufacturing arm, the Northern Electric Company. Northern Electric was permitted to manufacture domestic broadcast receivers as well as transmitters.

There were many independent telephone companies in the begin-

1923: CKNC—North Bay, Ontario, Canada (Photo courtesy Public Archives of Canada)

Northern Electric radio broadcast transmitting equipment. The No. 3-A is a 50-watt transmitter for use in point-to-point communications work for lumber camps, forestry preservation, police outposts, aircraft services, and emergency work.

ning days of broadcasting. One of these that was involved in receiver manufacture in Toronto was the Canadian Independent Telephone Co. It was formed to manufacture telephone equipment for the independent telephone operating companies, and in about 1922 began to manufacture broadcast receivers. The company was taken over by Bell Canada in 1924. Ted Rogers began his commercial radio career at the Canadian Independent Telephone Co. When the company was dismantled, its broadcasting assets were acquired by the Rogers family, as we shall see in the next chapter.

Manufacture of Broadcast Receivers by the Northern Electric Company in the 1920's

Robert P. Murray

The history of the telephone and of radio are connected in several ways. The Bell System fought off competition and emerged as the dominant holder of telephone patents in the latter 1800's. They had a serious interest in the technology of audio circuits, acquiring, for example, de Forest's audion patents. When broadcasting began in the early 1920's, the Bell System became a major player in the market for broadcast transmitters. Others were well positioned to produce broadcast receivers, most notably R.C.A., the result of a merger between the American Marconi Company and General Electric.[1] One result of the radio group patent negotiations in 1921 was that AT&T, or more specifically Western Electric, was to continue making broadcast transmitting equipment but was prevented from making domestic broadcast receivers.

In Canada there was a similar radio patent pool, although it came a couple of years later. Most of the players were subsidiaries of the U.S. companies in the business of making broadcast equipment, but events took a different turn. Specifically with respect to the telephone industry, the Bell Telephone Company of Canada, Ltd., through their subsidiary Northern Electric would continue to provide broadcast transmitters, *and* they were able to manufacture home broadcast receivers.

Why was there this difference between Canada and the U.S. in the manufacturing environment? One possibility, although I have never seen it documented, is that the U.S. radio group was created close to the beginning of broadcasting in both countries, in 1921. Despite accounts of the forward thinking of David Sarnoff at R.C.A., and his prediction of the Radio Music Box[1], no one had yet demonstrated the future potential of broadcasting in 1921. On the other hand, the Canadian patent pool dates from 1923, a year or more after broadcasting had begun. It is more likely that by then, AT&T would have seen a glimmer of the vast enterprise that broadcasting would become, and chose to negotiate more firmly for an unfettered position in it for its subsidiary Northern Electric.

But I am getting ahead of myself. Several details in the history of the telephone companies are where this story begins.

Background

Looking back, Melville Bell, Alexander Graham Bell's father in Brantford, Ontario, was assigned the patent rights to the telephone in Canada, but he had neither the knowledge nor the interest to manage a telephone company.[2] It might even be fair to say he didn't want a telephone company. Alexander Graham Bell had first demonstrated the electrical transmission of intelligible speech in 1876. Already at that time, he was using the production capability and shop space of Charles Williams Jr. in Boston to manufacture his inventions (not to mention the help of Williams' employee Charles Watson, who had become Bell's assistant in 1875).[3]

The telephone as a business had its beginnings in 1877, and that same year Melville Bell advertised telephone service in Brantford. In order to do this, he signed an agreement with Charles Williams Jr. of Boston, to deliver 1,000 hand telephones in exchange for the assignment of one-quarter of the patent in Canada. Melville Bell then appointed agents across Canada and launched the business of renting telephones. His efforts were not noticeably successful, because rights of way along highways could seldom be secured.[4] In 1879 Melville Bell sold his interests in the Canadian Bell patents to W.H. Forbes of Boston, who was president of the National Bell Company, the result of a merger of New England Telephone and Bell Telephone in the United States.[4]

Charles Williams Jr. was unable to supply the 1,000 telephones to Melville Bell in Canada because of the huge demand in the United States. He therefore hired James Cowherd, a tinsmith in Brantford, to be his Canadian agent. Cowherd had helped Alexander Graham Bell make the original prototype of his invention. Cowherd successfully operated a factory at

Brantford, and by October 1880 he had supplied 2,398 telephones to the Canadian company. Unfortunately, he died from tuberculosis on February 22, 1881.

The Western Electric Company was a subsidiary of Western Union, and quickly outclassed the small Bell and Williams manufacturing companies, turning out superior equipment. When Alex Bell agreed to an out of court settlement with Western Union in November of 1879 to the effect that Western Union would transfer its telephony business to Bell, and Bell would stay out of the telegraphy business, Western Electric became a subsidiary of Bell.[3] (Chap. 4)

In 1880 Charles F. Sise was sent by National Bell to bring some order to the telephone industry in Canada. He obtained a charter and organized the Bell Telephone Company of Canada in May of that year. The factory at Brantford, initiated when Melville Bell owned the Canadian Telephone patent, was closed. From 1880 to 1882 repair shops at Montreal and Hamilton also did small manufacturing for the Company. During this period, the larger items of machinery were purchased from independent manufacturers in Canada or the U.S.A.[4]

The first Bell Telephone factory and repair shop was opened as the Mechanical Department in Montreal in 1882. As the Company was determined to manufacture in Canada, a new company was incorporated for the purpose in 1895, under the name of the Northern Electric and Manufacturing Company, Limited. The new company was granted broad powers, which would come to have significance at a later time. An excerpt from the letters patent incorporating the company reads:

> (a) To manufacture and deal in brass, copper, and other metals and wires, rods, cables, lamps, castings and wrought metal articles, and their accessories; (b) To construct or contract for the construction for others of electric light, power, telegraph, telephone or street railway cable lines or plant and appliances or articles used in connection therewith; (c) To own, use and operate one or more line or lines of telegraphic or telephonic communication, and to purchase or lease electric light, telegraph or telephone plant, works or apparatus; (d) to acquire stock or shares in any electric light, telegraph, or metal wire, brass or street railway cable company as the consideration for goods, wares, or merchandise sold by the company to such other company in the ordinary course of business; (e) To acquire such licenses, patents or industrial designs as may be deemed necessary or expedient for the purposes of the said business and to alienate the same at pleasure.[5]

A larger factory was built equipped with more machinery to meet the growing demands of the Telephone Company. While machinery was installed in order that the equipment might be economically produced, the demand was not sufficient to keep the new machines fully utilized. Therefore it was decided to take profitable contracts for other than electrical apparatus.[6] In 1914 this company was merged with a wire and cable company also owned by Bell Telephone of Canada, and renamed Northern Electric Company, Limited.[4]

The history of stock ownership of the Northern Electric and Manufacturing Company, followed by Northern Electric Company, Limited, is shown in Table 1. For most of its history, the Bell Telephone Company of Canada owned the majority of NE stock, and Western Electric Company held a minority position. This relationship held until 1957, when a court decision required AT&T, and hence Western Electric, to divest its Canadian holdings.

Table 1. Northern Electric Company Limited, History of Stock Ownership

1. The Northern Electric and Manufacturing Company Limited (Incorporated 1895)			
Year	Bell Telephone Co.	Western Electric Co.	Others
1895			100%
1896	93%		7%
1906	55.80%	40.00%	4.20%
1910	55.20%	40.00%	4.80%
1911	50.00%	45.20%	4.80%
2. Northern Electric Company Limited (Incorporated 1914)			
1914	50.00%	43.60%	6.40%
1929	56.31%	43.57%	0.12%
1957	89.97%	10.52%	0.01%
1962	99.99%	0.01%	
1964	100.00%		
Northern Electric Company, Ltd. Secretary's Department, August 17, 1967.			

Patents were suspended in Canada for the duration of the First World War, 1914-18. Having contracts for war production, the electric companies were accustomed for a time to manufacture equipment using all known patents, as needed.

On August 14, 1923, the Toronto Star announced that six electric companies had agreed to pool all their patents in order to avoid litigation and improve radio. This step was taken following the example of a number of companies in the United States in 1921, that had merged their radio-related patents with those of R.C.A.. The companies involved in Canada were the Canadian General Electric Co., the Marconi Wireless Telegraph Co. of Canada, the Canadian Westinghouse Co., the Bell Telephone Co., the Northern Electric Co., and the International Western Electric Co.[7] Under the terms of the agreement, each company agreed to license the other parties within their natural fields.

In a 1924 newspaper article, the relationship between the telephone company and the manufacturer was explained as follows.[8] The Bell Telephone Co. in Canada began to appreciate early on that it required sophisticated and well machined apparatus that would be reliable in service. They further realized that no such manufacturing business, familiar with telephone apparatus, existed in Canada. Catering to the needs of almost all of the telephone companies in Canada, it became apparent that economies of scale could be achieved if Northern Electric (NE) supplied all of the needs of Bell Telephone, and manufactured and stored all such equipment centrally. The manufacturing company also agreed to purchase and store goods not of its own manufacture, and to sell these goods on the open market where opportunities existed. In their 1925 general catalogue, Northern lists for sale telephone equipment, signal and message systems, wires and cables, household appliances, illuminating equipment, wiring devices and electrical supplies (the largest section, and includes fire extinguishers), line construction material and tools, and power apparatus (anyone need a 200 hp electric motor?).[9]

They also supplied radio broadcasting equipment, of course. In reference to a 50 watt transmitter, they say, "In order to meet the ever-increasing demand for a small power radio transmitting equipment for use in the woods, the Northern Electric Company is offering …" [9 (p. 39)] While sitting in the woods writing this article, it occurred to me that just maybe the advertiser had creative assistance from someone who lived in a more populated country.

Radio was not the core business at NE, but still a view of the financial information (Table 2), indicates that it was a substantial business. By 1924, it had grown to represent more than 9% of sales. (The years in Table 2 are not, strictly speaking, comparable. However, after 75 years have elapsed, it is not possible to call the Finance Department and ask for the missing pages.)

It should also be recognized that the NE radio business included the manufacture of broadcast transmitters and the sale of transmitters made by the Western Electric Company. No information remains that treats the transmitter and receiver businesses separately, but evidently NE held a strong position in the broadcast transmitter market.

What can we infer from this table of financial information? Although gross profits for 1921 to 1923 are apparently calculated differently from those of 1924, and although net profits in 1924 appear to

Table 2. Northern Electric Company, Limited Sales and Gross Profits by Year by Class of Merchandise (a)

Year	Telephone Apparatus	%	Wire and Cable	%	Radio	%	Merchandise Not N.E.	%	Total	Radio % of Total
1921										
Sales	4,721,017		2,752,996				4,263,373		11,737,386	
Gross Profits	1,143,000	24.2	228,000	8.3			394,898	9.3	1,765,898	
1922										
Sales	4,169,181		3,478,451		228,194		3,771,342		11,647,168	1.96%
Gross Profits	914,590	21.9	522,193	15.0	(b)		454,766	12.1	1,891,549	
1923										
Sales	5,600,290		5,279,738		719,048		4,585,016		16,184,092	4.44%
Gross Profits	1,369,465	24.5	846,514	16.0	245,378	34.1	569,053	12.4	3,030,410	8.10%
1924 (c)										
Sales	939,778		1,826,506		530,412		2,327,894		5,624,590	9.43%
Gross Profits	152,080	16.2	185,629	10.2	131,404	24.8	371,440	16.0	840,553	15.63%
Expenses	115,388		143,203		70,455		238,453		567,499	12.41%
Net Profits	36,692	3.9	42,426	2.3	60,949	11.5	132,987	5.7	273,054	22.32%

(a) For the years 1921 to 1923, this information appears to apply to the whole company.
(b) Gross profits included with telephone apparatus.
(c) For the year 1924, the available information applies only to the sales regions. Expenses in the sales regions include the categories Administration & Rent, Stores Expense, Sales Expense, and Financial Expense, so they are unrelated to the costs of production. Therefore net profits are not available for the whole company, and indeed may not have been available at the time.

apply only to regional sales, we can still draw an overall generalization. Looking horizontally across the page at each definition of profit separately, it appears that radio was NE's most profitable area of business in each instance.

It is curious, then, that they abandoned the domestic receiver business altogether in 1926, as we shall later see. It is a year for which we have no corresponding financial data.

Being as diverse a manufacturer as it was, the initial move into broadcast receivers must have seemed normal. In the Company's words, "With the advent of radio it was but natural that the Northern Electric, who have specialized in the reproduction of the human voice for over a quarter of a century, should be prepared to develop and perfect radio apparatus" (NE sales brochure, 1924). What did they develop? A list of the early models, produced with the obvious help of the Western Electric Company, is shown in Table 3. All of these are battery sets.

Domestic Receivers

The Model R-1 appears to have been the first broadcast receiver produced by NE. That is to say, its model number and the fact that it was introduced in the first of a series of Northern Electric Radio Bulletins (No. 1000, Nov. 1, 1922) suggest that the R-1 was the earliest. NE did not first produce crystal receivers in Canada, as the Canadian Marconi Company and the Canadian General Electric Company did.

The Model R-1 is shown in Figure 1, along with the R-5 and R-105 amplifiers, and the R-20 variometer sitting on top of the receiver. The R-1 advertised in Radio Bulletin No. 1000 was not equipped to accommodate a variometer. Later models of R-1, including the one pictured here, came with two terminals on the front pan-

Table 3. The broadcast receiver models of the Northern Electric Company Ltd. in the 1920's.

MODEL	YEAR	CIRCUIT	TUBES	STAGES	CABINET	PRICE
R-1	1922	Det.	1 (R215-A)	Det.	Box	30.00
R-2	1922	TRF	3 (R215-A)	2RF-Det.	Box	85.00
R-3	1923	TRF	4 (R215-A)	2RF-Det.-A	Table	100.00
R-4	1924	SH	6 (R215-A)	SH	Table	205.00
R-4-L	1926	SH	6 (R215-A) & 1 (R-221-DX)	SH	Table	
R-5/R-5A	1922	Amp.	1 (R215-A)	A.	Box	22.50
R-11	1923	Reg. Det.	1 (R215-A)	Det.	Box	40.00
R-12	1923	Reg. Det.	2 (R215-A)	Det.-A	Box	
R-15	1923	Amp.	1 (R215-A)	A.	Box	22.50
R-20	1923	Variometer	-		Box	10.00
Victor Northern Electric Models						
R-20	1925	Reg. Det.	2 (R215-A)	Det.-A.	Table	42.00
R-21	1925	Reg, Det.	3 (R215-A)	Det.-AA	Table	68.00
R-22	1925	Reg. Det.	3 (R215-A)	Det.-AA	Victrola	54.00
R-23	1925	Reg. Det.	3 (R215-A)	Det.-AA	Victrola	64.00
R-24	1925	Reg. Det.	3 (R215-A)	Det.-AA	Table	
R-30	1925	TRF	5 (R221-DX)	TRF	Table	175.00
R-31	1925	TRF	5 (R221-DX)	TRF	Victrola	135.00
R-40	1926	SH	7 (R215-A)	SH	Table	225.00
R-41	1926	SH	8 (R215-A)	SH	Table	260.00
R-50	1925	TRF	5 (3-R215-A & 2-R221-DX)	TRF	Table	125.00
Northern Electric Power Amplifiers						
R-105	1923	Amp.	1 (R208-A)	Amp.	Box	30.00
R-106	1923	Amp.	2 (R208-A)	AA	Box	50.00
R-107	1923	Amp.	3 (R208-A)	AA	Box	70.00

Note: NE models are priced without tubes. Models offered jointly with the Victor Talking Machine Co. are priced with the tubes included.

el which allowed the user to interrupt the plate circuit of the single tube with the variometer. Sitting above the antenna transformer, the variometer provided enough adjustable feedback to convert the R-1 to a regenerative set. Without the variometer, (and with a nickel plated jumper to replace it), the R-1 would have been a feeble performer. More will be said about the amplifiers later. The Model R-1, in combination with two R-5 amplifiers, tubes, batteries, phones and antenna equipment was advertised as the R-1000 Radio Receiving Set for the price of $105.00.

Fig. 1. Northern Electric R-1 receiver, R-5 and R-105 amplifiers, and R-20 variometer

I found it curious that a manufacturer that had access to the Armstrong regeneration patent, from 1923 at least, would choose to offer a variometer for sale that would make their set regenerate but would not provide a complete regenerative set. Ralph Williams, in describing the Atwater Kent company, offers some suggestions on this topic.[10] He points out that regenerative sets were not easy to tune, and required some working knowledge of the adjustment procedure. Secondly, regenerative sets tended to go into oscillation, which became a major disadvantage as receivers became popular. The receiver transmitted an annoying signal, close in frequency to the desired station. Finally, he observes that a regenerative set was not effective enough to survive as a family radio, although a patient radio enthusiast might make one behave well. For these reasons it was not in the interest of a manufacturer who sought to promote high quality radio receivers to include a regenerative model as a complete set.

The business side of the receiver panel is shown in Figure 2. The coils of the antenna transformer are along the top. The R-215A tube and the grid leak are fastened to what appears to be a zinc plated steel shelf. The wiring is done with rubber covered tinned 20 gauge solid wire, routed at least part of the time at right angles like buss wiring. The wiring of this set appears to be original, except for some mouse-made amendments to the insulation. At this point in time, NE acted as though it were manufacturing equipment for rent rather than sale, and they conveniently pasted schematics inside their cabinets just as they did with early telephones (Figure 3).

Fig. 2. Northern Electric R-1 components side of front panel

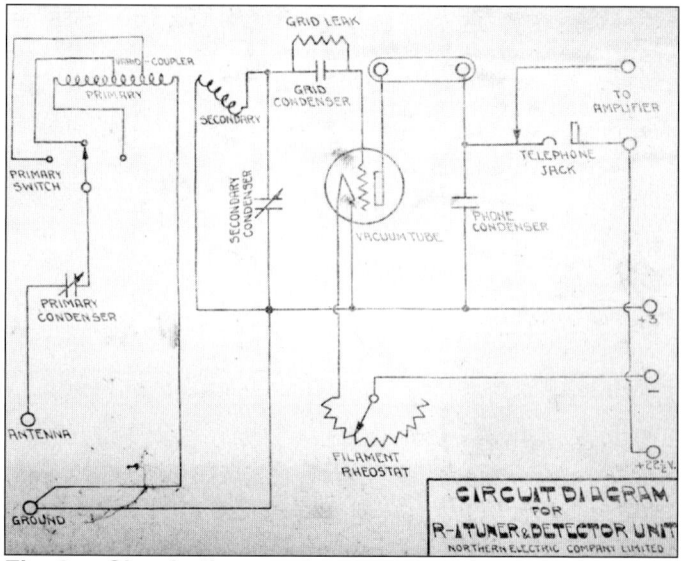

Fig. 3. Circuit diagram for the R-1

A month later, in December 1922, the R-2 receiver was launched. Its period of development probably overlapped considerably with that of the R1. It consists of a tuning circuit followed by two stages of rf amplification and a detector/audio amplifier. The controls (Figure 4), in addition to the filament rheostat in the upper center of the panel, are (left to right) the primary condenser (PC), the primary inductance switch (PI), the coupler (C), the stabilizing rheostat (S), and the secondary condenser (SC). The circuit is designed much like those of crystal sets of the time, with the primary condenser and primary inductance connected in series and serving to tune the set. The loading inductance adjusts the coupling of the antenna circuit to the receiver (Figure 5). There is no mention of feedback in the operating instructions, and I have not attempted to operate one myself. The stabilizing rheostat and the coils are not original in the set in Figure 6. Why is it that when a young family member takes an interest in an obsolete set, the coils are often the first

Fig. 4. R-2 Short Wave Tuner and 2 Stage Radio Frequency Amplifier

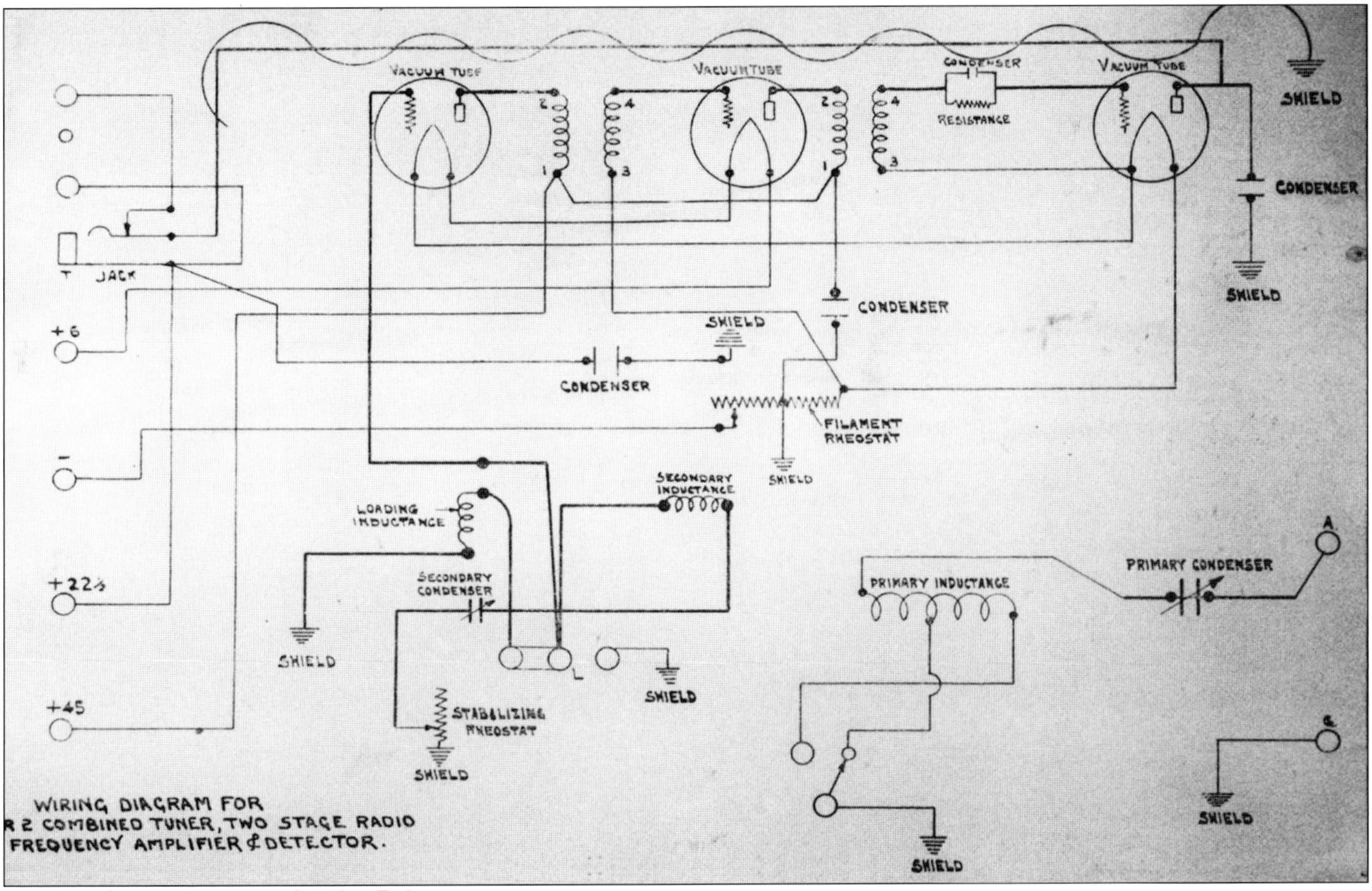

Fig. 5. Circuit diagram for the R-2

component to go? The youthful interest may save the set, mind you, from the town dump. The coils in this set were restored with information obtained at the Hammond Museum, Guelph, Ontario, Canada.

Western Electric obviously had a hand in designing this receiver, although at the time they were not producing any domestic broadcast receivers of their own. The NE description says, "The two stages of radio-frequency amplification embodied in this unit employ radio-frequency transformers specially designed by the engineers of the Western Electric Company, New York, for this purpose." These are labeled "1" and "2" on the tube shelf in Figure 6. The wiring in this receiver has progressed from the rubber covered solid wire to stranded, fabric covered hook-up wire. The Model R-2, along with a R-5A amplifier, phones, tubes, batteries and antenna equipment was also marketed as the R2001 Radio-Frequency Receiving Set, price $150.00.

The July, 1923 Radio Parts Catalogue makes reference to the R-3 receiver (Figure 7), and the August 1, 1924 Radio Bulletin describes it in more detail. At this stage in development, they have managed to omit some of the controls from the panel. NE has also ceased the practice of pasting a schematic diagram of the circuit into the back of the cabinet. The set appears to be an improved model R-2. It has two stages of r.f. amplification, followed by a detector and an additional stage of audio amplification. It has fewer controls than the R-2. Neither model R-2 or R-3 had adopted the TRF design with three tuned stages cascaded after one another. The R-3 shares some similarities with the Western Electric 3B.[11] It's internal parts are shown in Figure 8. Compared to the R-2, the R-3 has adopted tuning condensers with gear-driven vernier controls, but the coil arrangement is much the same. The r.f. transformers have been redesigned, and an audio transformer has been added to couple the extra stage. The front panel markings are all done with transfer decals applied to the painted wood panel.

The R-4 superheterodyne was introduced in August of 1924. By this time NE had shared the benefit of the Canadian patent pool for about a year. Because of the increasing number of broadcast stations and their locations closer to each other on the broadcast band, radio designs with increased selectivity were required. The superheterodyne design was a logical answer. The R-4 is shown in Figure 9. It has a somewhat fancier cabinet than its predecessors, but the similarities are also recognizable.

Fig. 6. R-2 components side of panel. The coils and the stabilizing rheostat marked "S" are not original.

Fig. 7. R-3 Four-tube Radio-frequency Set

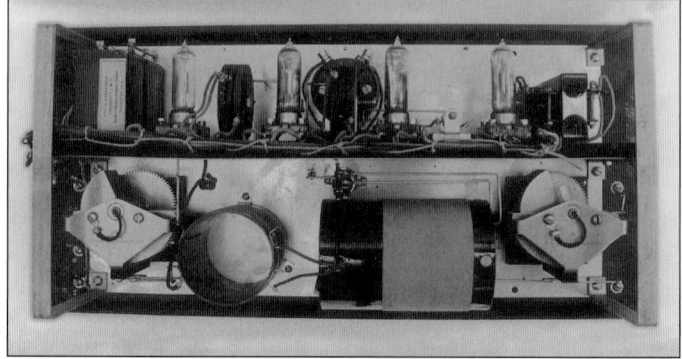

Fig. 8. R3 components side of panel

Fig. 9. R-4 Superheterodyne Radio Receiving Set

Although previous sets have been equipped to operate with a loop antenna, a loop was not recommended. The R-4 is the first in the series for which NE feels sufficiently confident to recommend a loop, and offer one for sale. The inside of the receiver is shown in Figure 10. In this set, heavy nickel plated brass shielding separates the oscillator from the tuning section, and both from the rest of the set. The wooden box is shielded as well. The NE R-4 is quite similar to the Western Electric 4D.[11]

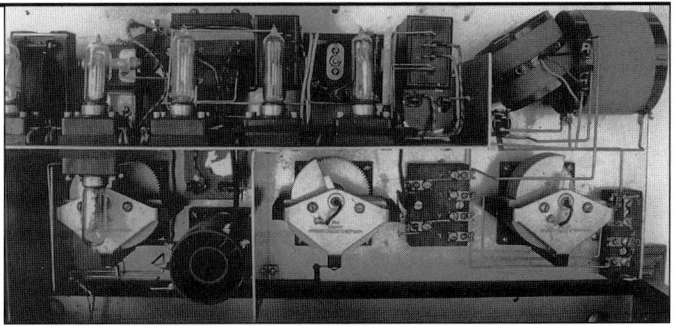

Fig. 10. Inside the R-4 receiver

Although NE had discontinued the supply of receiver schematics with its sets in 1922, there is an exception for the R-4. One of their engineers, W.B. Cartmel, published an in depth description of the set, including both a sketch of the receiver and a schematic diagram (Figure 11) in the January, 1925, issue of *Radio News of Canada*, a prominent radio magazine of the time.[12] The article runs to five full pages, only it doesn't mention the Northern Electric Company in relation to the receiver or Mr. Cartmel. He attributes the superheterodyne inven-

Fig. 12. R-11 Radio Receiving Set and R-15 Amplifier

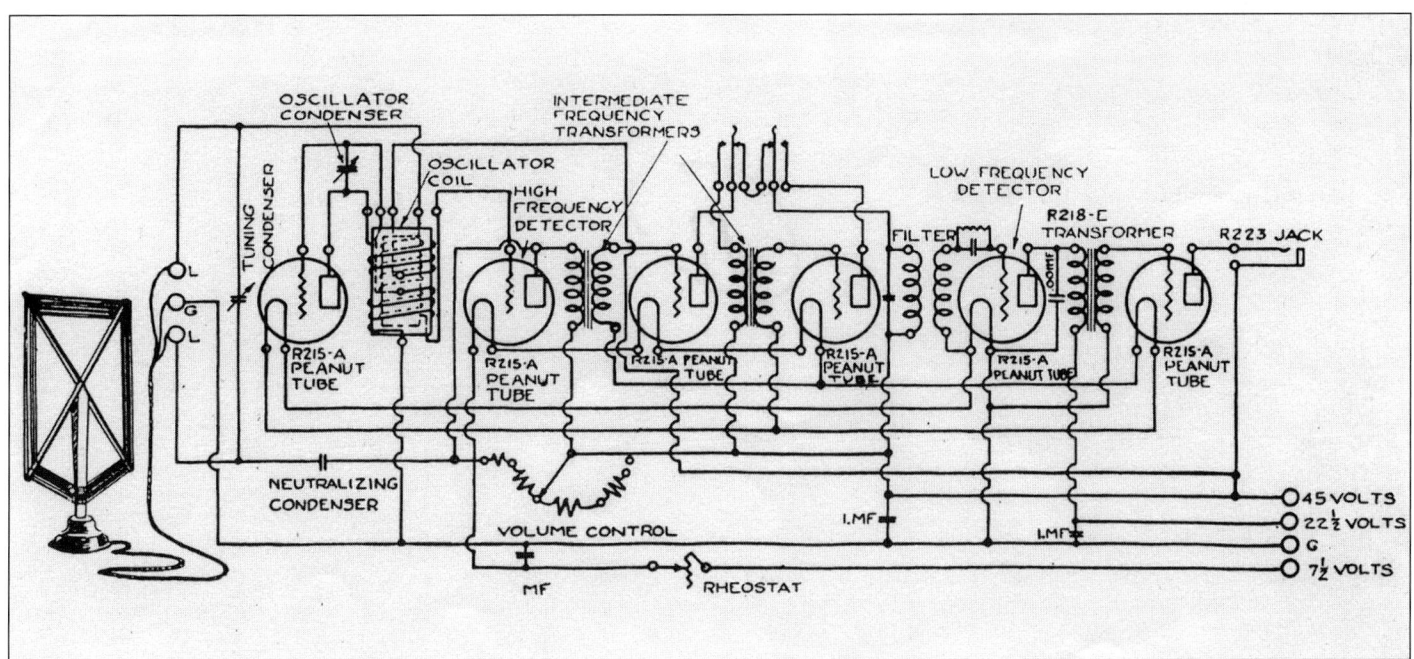

Fig. 11. Circuit diagram of the R-4 receiver

tion to Lucien Levy, without ever mentioning Edwin Armstrong. It is difficult to assess now in retrospect whether publication of the details of the R-4 represents an act of defiance on the part of Mr. Cartmel, or a decision by the company to disseminate this information.

The receiver has an intermediate frequency around 50 KHz, typical for the time, and this specimen at least has several images along the dial of the same station. NE advertising asserts that this does not happen. The set uses metal cored transformers at the intermediate frequency. The oscillator was tuned manually.

There was a model R-4-L produced later, which had the same external appearance, but with another stage of audio amplification contained in the back of the box. It used a R-221-DX tube as the power tube.

The model R-11/R-15 combination was offered in 1923 (Figure 12). It was a simplified equivalent to the R-1 but somewhat less expensively made. It had only three controls, tuning, feedback, and rheostat. Mind you, in August, 1924, the R-1 listed for $30.00 and the R-11 listed for $40.00. It is the first NE receiver described as "regenerative" in its original advertising, presumably because it was designed late in 1923, after the appearance of the Canadian patent pool. Its construction was not confined to NE parts. Some of these can be seen in the inside views of the receiver and amplifier (Figures 13 and 14). There was no schematic diagram provided in the back of the case.

Model R-12 was another in the same series. It consisted of a one-tube regenerative detector with one stage of audio amplification included on the same chassis. A battery compartment was located at the rear of the cabinet (Fig. 15).

Northern Electric Power Amplifiers

At the time it produced its first broadcast receiver the R-1, NE also offered the R-5 amplifier to those customers who wanted a somewhat stronger signal in their headphones. These audio amplifiers used the R-215-A tube (Figures 1 and 16), and could be added singly or in pairs to the output of a NE R-1 or R-2 set. The schematic diagram from the back of the case is shown in Figure 17. The rheostat in the R-5 amplifier was of a suitable value such that approximately 1.1 volts could be supplied to the filament when the A battery provided 3 volts. Later receivers (R-2 and upwards) used a 6 volt filament battery, and the R-5-A

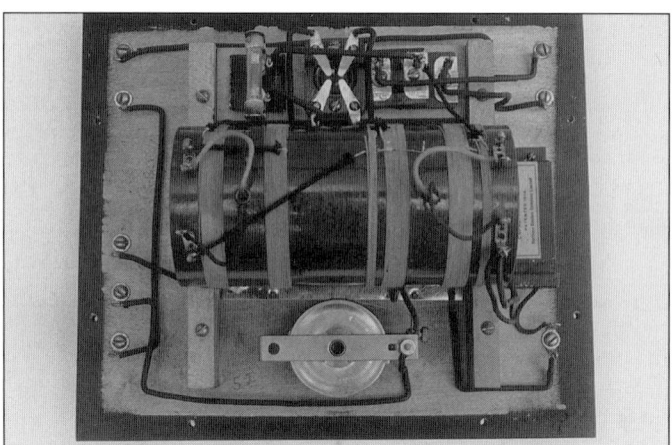

Fig. 13.   R-11 components side of panel

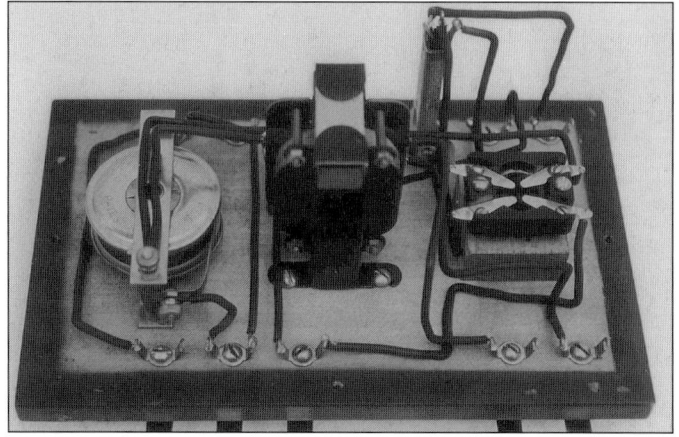

Fig. 14.   R-15 components side of panel

Fig. 15.   R-12 Radio Receiving Set

Fig. 16.   Components side of the R-5 amplifier

amplifier was fitted with a rheostat suitable for dropping this voltage to approximately 1.1 volts.

In 1923, still early in the sequence of NE receivers, a range of power amplifiers was offered to customers wanting loudspeaker performance from their receiver. The R-105 amplifier came in the same size box as the R-5, but contained a higher powered tube, initially the R-208-A and soon afterward the R-216-A. The R-105 is shown as the final amplifier in Figure 1. The component side of the R-105 is shown in Figure 18. The schematic diagram is in Figure 19.

At the same time the model R-106 was introduced (Figure 20). Again this was first offered with two R-208-A tubes, but was soon changed to R-216-A tubes, one tube acting as driver and the other as final amplifier. There was a shelf on which a 9 volt C battery could be fastened (Figure 21). Volume control

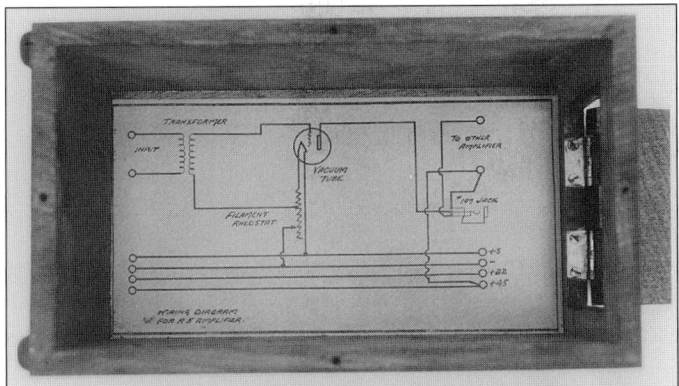

Fig. 17. Schematic diagram of the R-5 amplifier

Fig. 20. The R-106 Loud Speaker Amplifier. Dial plates on this amplifier are reproductions.

Fig. 18. Components side of the R-105 amplifier

Fig. 21. Components side of the R-106 amplifier. Some of the wiring is not original.

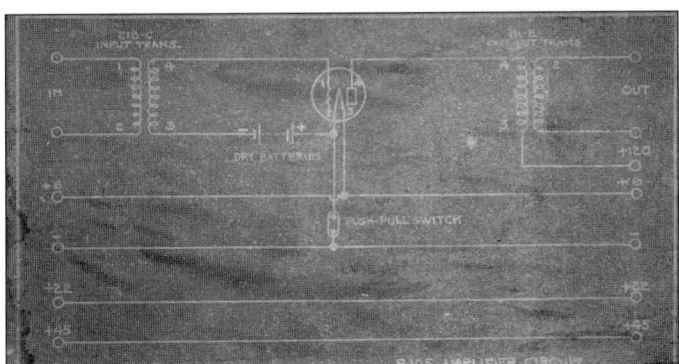

Fig. 19. Schematic diagram of the R-105 amplifier

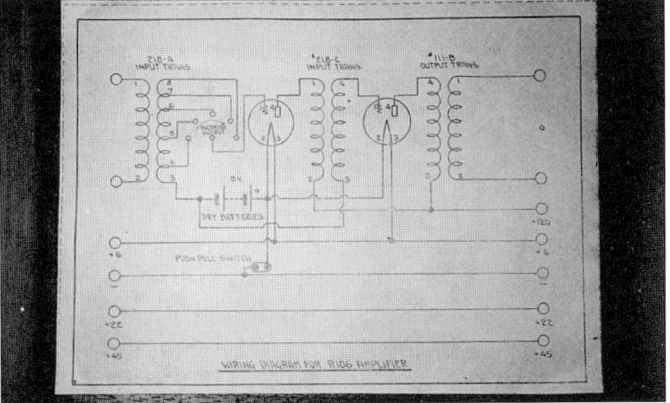

Fig. 22. Schematic diagram of the R-106 amplifier

was achieved by a tap switch selecting from five points on the input transformer, (Figure 22).

The model R-107 amplifier is a two-stage audio amplifier with three R-216-A tubes, (originally R-208-As), acting as driver and push-pull stages (Figure 23). This is the NE version of the Western Electric 7A amplifier. There are differences in both layout and cabinet design, but the circuit appears to be identical between the two.[11] There was no filament control in these amplifiers, since the tubes they were using worked well at a full 6 volts the manufacturer saw no need for one. End users, however, often added one as an "owner improvement." The R-107, like the R-106, also controls volume by selecting taps on the input transformer (Figure 24).

Victor/Northern Electric Receivers

Cooperation between NE and the antecedents of the Victor Talking Machine Company go back at least to February 12, 1900, when C.F. Sise, then president of NE, noted in his log book that he had "Offered to Berliner, for the manufacture of his gramophone, 3 horsepower, room 15' x 17.5', heat and light, at $45.00 per month."[13] In 1900, the first gramophones in Canada were manufactured in space provided by NE in its Aqueduct Street building. The inventor was Emile Berliner, inventor of the flat disc records that were to replace Edison cylinders and become the industry standard. Berliner was an immigrant from Germany to the US in 1870, and lived and worked in Washington and Philadelphia. He registered several patents, and formed the United States Gramophone Company in 1893, and then the Berliner Gramophone Company in Philadelphia in 1895. The latter company was sued by the former over title to his inventions in 1900. As a result, Berliner folded the Berliner Company in the US and moved his operations to Montreal.[14]

The NE company also rented space to Berliner for his pressing plant which manufactured flat disc records.[13] The Berliner Gram-o-phone Company built a factory of its own in 1908, and another in 1912, both on Lenoir Street in Montreal.

In 1924, Berliner Gram-o-phone was bought by the Victor Talking Machine Company of Camden, NJ. In 1929 Victor joined with RCA to become RCA Victor.[14]

Some other exerpts from the log books of C.F. Sise, president of NE:[13]

February 25, 1920 – NE employees' Pension and Benefit Plan excluded claims resulting from fooling around, violent or riotous or immoral living.

September 22, 1920 – Alliance between Canadian General Electric Co. Ltd. and Marconi Wireless Telegraph Co. of Canada Ltd. and the Canadian Radio Corporation "no great fear of competition from the group ..."

December, 1922 – NE began manufacture of R2001 receiver set (higher amplification than R1000); plus short wave tuner (R2) and detector unit.

January 1923 – CHYC – 500-watt transmitter installed on 7th floor Shearer St. (NE broadcast station).

April 1, 1923 – CHYC – first church broadcast.

April 24, 1923 - $3,420 appropriated for one only single six Packard touring car, seven passengers.

July 24, 1923 – NE appropriation for tools for mfr of R3 and R11 radio receiving sets; and R15, R105, R106, R107 amplifiers.

April 23, 1925 – NE to execute contract with Victor Talking Machine Company of Canada Limited for handling by Victor of exclusive sales of NE radio receiving sets in Canada.

March 1, 1929 – Broadcasting station sold to La Presse Ltd. (CKAC) for $89,000 on condition NE takes a certain amount of time; NE to use small 100-watt set to broadcast church services until new La Presse station operational.

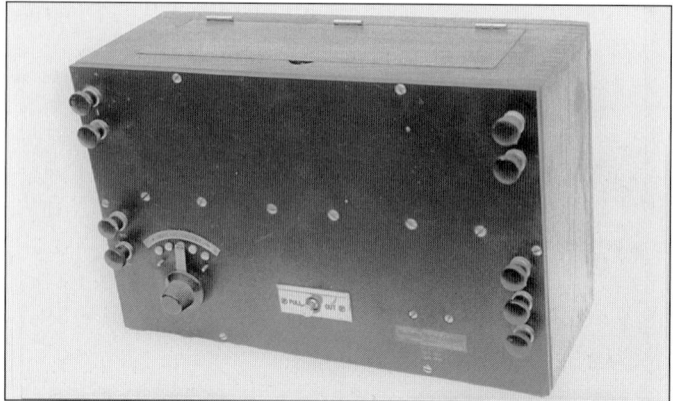

Fig. 23. The R-107 Loudspeaker Amplifier

Fig. 24. The components side of the R-107 amplifier. Some of the wiring is not original.

On April 23, 1925, NE reached an agreement with the Victor Talking Machine Company of Canada Limited of Montreal. They began production of a series of receivers to which NE apparently contributed the chassis and Victor the cabinet. Electronically these radios represented no particular advance for NE, which had by now been producing superheterodynes (the R-4). The first of these Victor/Northern Electric models was the R-20, which was a regenerative detector and amplifier, a set using two R-215-A tubes and other NE parts. It was much like the R-12, and is shown in Figure 25. Like the R-12 it was simple to operate, having only three controls. The components side is shown in Figure 26. It looks rather like a Radiola III, don't you think? Its cabinet was somewhat more elaborate. It was described as having enough space inside to accommodate the batteries. A schematic was not supplied.

The Victor/Northern Electric type R-21 is a similar set using three R-215-A tubes in a regenerative circuit with two stages of audio amplification. Like the R-20, it claimed room inside the cabinet for the batteries, a little more convincingly this time because the cabinet was almost twice as long as the radio panel. A version of the R-21 was provided as a horizontal radio panel in a Victorola gramophone cabinet and was labeled type R-22. The same receiver provided as a vertical panel in an upright Victorola was labeled type R-23.

The Victor/Northern Electric type R-24 was again a three R-215-A tube regenerative set with two stages of audio amplification. This time it was in a slightly fancier box made of American Walnut (Figure 27). The components are shown in Figure 28. Again, it has Radiola III style coils. This was also the first broadcast receiver with which NE was involved, that did not display the tubes through holes in the front panel. The market seemed to be getting a bit more sophisticated in its tastes. Type R-30 was a TRF receiver of the 1925 model year. It was a typical looking "three-dialer" sharing some of the styling of the R-20 series of sets. It used 5 of the R-221-DX tubes, which were functionally equivalent to the ubiquitous R.C.A. UX-201A. Type R-31 was the horizontal panel version of the same set, designed to reside inside a Victorola cabinet.

Following the R-4 and R-4-L superheterodynes, NE produced nine models with simpler circuits before attempting another superheterodyne. That is, if they used model numbers sequentially which seems to have been approximately true. The question is why?

Fig. 25. Victor/Northern Electric type R-20 receiver

Fig. 27. Victor/Northern Electric type R-24 receiver

Fig. 26. Components of the R-20 receiver

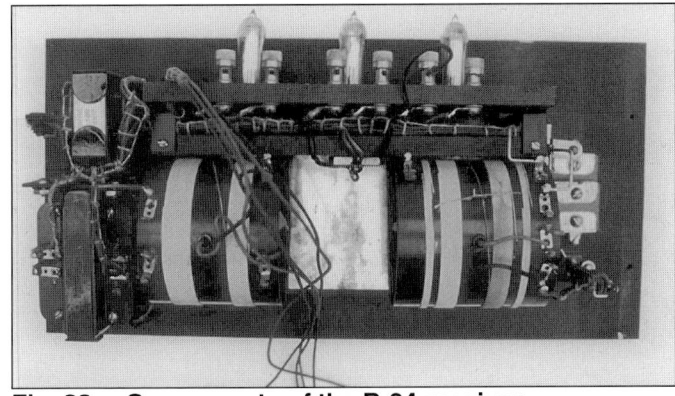

Fig. 28. Components of the R-24 receiver

Fig. 29. Victor/Northern Electric type R-50 receiver

Fig. 30. Chassis of the R-50 receiver

No one can tell us, 70 years later. The R-4 superheterodyne appears to have sold quite well, judging by the number of surviving specimens. It was certainly expensive enough in comparison to its contemporary models. It seems possible that when reviewing the sales figures from 1922 to 1924, NE management decided to emphasize less expensive sets. However all of these models were bad neighbors in the sense of transmitting howls when not tuned carefully. Not since the R-3 had they protected the antenna against transmitting oscillations with a stage or two of r.f. amplification. The discretionary dollars would have also been scarcer in Canada. Oh well, we were in the woods, remember?

Finally, in 1926 came another superheterodyne., the type R-40. This set used seven tubes and was styled very much like the type R-30 and the R-24 before it. For a little more money there was the type R-41, an eight tube superheterodyne, apparently in the same cabinet. My bet is on an extra stage of audio amplification (see Table 3 for list prices and dates).

In 1925 there had been another anomaly, the type R-50. This was a five tube set, a "three dialer", using three R-215-A tubes and two R-221-DX audio tubes (Figure 29). The R-215-A tubes and the R-221-DX tubes are alternated in a single row inside the set (Figure 30).

After 1926, NE and Victor appear to have gone their separate ways. The Victor Talking Machine Company made receivers under their own name for a time, and eventually presented models under the RCA Victor name, models which were the result of the merge of the parent companies in the United States in 1929. At NE, this point marked the end of receivers that were obviously designed by telephone company engineers and manufactured by the telephone industry. After spending a few seasons in disarray, they began to market receivers built by other companies, most notably American Bosch. The American Bosch model 5A became the NE Minaki (1931) and the American Bosch 200-A became the NE 51 (1932) for example. NE later resumed the production of radios in a style which by then had become much closer to that of the rest of the industry. They used generically available parts and mass production methods.

Why did NE withdraw from the broadcast receiver market for the years between 1927 and 1930? One reason may have been that they lacked solutions to the electrical problems presented by alternating current (a.c.) operated sets. The telephone industry had progressed a little more slowly in solving these problems than had the broadcast radio industry. Western Electric had not begun to produce tubes with indirectly heated cathodes until 1929, and NE had, until that time, relied only on Bell System tubes.[15] The receivers that NE obtained from American Bosch in the early 1930's were a.c. operated, suggesting that indeed they saw this feature as responding to a demand of the market.

Northern Electric Radio Tubes

NE first offered to the radio market what they conceded were telephone repeater tubes. In an early radio parts catalog, (July, 1923), they listed two vacuum tubes for radio apparatus: the R-215-A and the R-208-A. The prefix "R" was not found on all NE tubes, and John Radcliffe (retired Curator, Bell Canada Historical Collection), suspects it may either have stood for "radio" or "retail". Although the R-203-D was advertised in an August, 1924 catalog, the four examples of this tube that I have seen are all marked R-203-B, Made in Canada, Northern Electric Co., Limited.

The R-208-A was apparently the first NE tube actually manufactured in Canada. What follows is a table of NE radio tubes sold in Canada (Table 4). Corresponding to Table 4 are Figures 31, 32 and 33. These tubes are either possibly or definitely designed by the Western Electric Company.

The R-215-A tube was widely used in domestic radios in Canada, both in factory built radios[17,18] and in the home builders market[19]. It is interesting because although the early development was attributed to H.J. van der Bijl, a South African radio pioneer working for Western Electric in New York[20], a popular Canadian radio magazine attributed it to W.B. Cartmel, a NE engsineer[21]. This "fact" appears to have escaped the attention of both Gerald Tyne[16] and John Stokes[22]. It is curious that Cartmel would have been represented this way. Although charlatans were certainly in existence at that time in Canada as elsewhere, the telephone industry was particularly subject to litigation, had their own legal department, and one presumes would not have looked favorably on "small" and unnecessary legal transgressions. Cartmel, also guilty of publication of the R-4 receiver schematic and design, sounds like a black sheep, at least.

Today there are numerous 215-A tubes in collectors' hands. The ones made by NE typically do not have the number R-215-A printed on them. They do

Table 4. Northern Electric tubes of the 1920's that were sold as radio receiving tubes

Number	Year	Use
R-203-D	1924	Developed for military use and s old surplus for amateur use after W.W. I. Listed in a NE radio parts catalog, 1924.
R-208-A	1922	First tube manufactured in Canada by NE for usein telephone repeaters (Feb. 3, 1922).
R-215-A	1922	Used extensively in the first NE receivers. Early development attributed to H.J. van der Bijl during W.W. I.
R-216-A	1922	Replaced the R-208-A. in NE power amplifiers in 1924.
R221-D or R-221-DX	1926	Equivalent to RCA UV-201A. Equivalent to RCA UX-201A, used in R-4-L, R-50 etc.
DX-235	1926	Equivalent to RCA UX-201A.

Note: The information in this table comes largely from Magers[15] and Tyne[16].

Fig. 31. R-203-B and R-208-A tubes

Fig. 32. R-215-A and R-216-A tubes

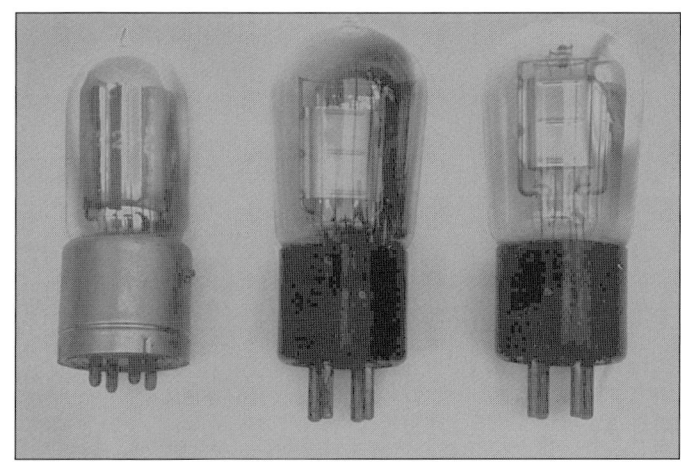

Fig. 33. R-221-D, R-221-DX and R-235-DX tubes

have a small circular NE logo molded into the bottom of the bakelite base between the button contacts (Figure 34). The Canadian 216-A is referred to as either the 216-A or the R-216-A. It may also be true that they were listed as R-216-A in the catalogs, but marked 216-A on the nickel plated brass base of the tube. That is how it is in my experience, but I can think of no reason why it should be so.

The NE type 221-D comes in two versions (Figure 33). First there is the R221-D which has a UV base and is the earlier of the two. NE apparently had no stock of metal bases stamped R221-D. The six that I have seen have all had the tube number ground off of the side of the base, and painted on the side of the glass. Judging from the patent dates remaining on the metal, the bases might have been originally intended for R-203-B tubes. The newer R-221-DX tubes have a UX bakelite base with tube numbers and patent dates molded into the bottom of it (Figure 35). The "Made in Canada" mark on these tubes was separately applied in silver paint, raising the possibility that they were made by Western Electric, rather than just designed by them. The DX-235 was another RCA UX-201A equivalent made about the same time, but it had a tipless bulb with the evacuation connection on the bottom of the tube (Figure 36). The DX-235 was advertised as a "Nor-Phonic" tube (Figure 37). There was not enough information supplied in the advertising to indicate what design improvements in the tube accounted for the advertising claims. The DX-235 originally sold for $2.00, but in 1929 they were considered surplus stock and were offered in the Eaton's department store radio catalog for $0.79, delivery included. The age of the 201-A was over.

Some of the early NE tube cartons are shown in

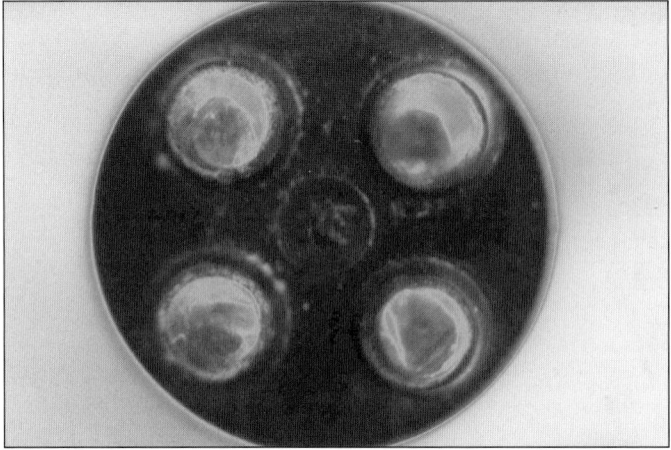

Fig. 34. Bottom view of an R-215-A tube by Northern Electric

Fig. 35. Bottom view of an R-221-DX

Fig. 36. Bottom view of a DX-235

Fig. 37. Print ad for a NorPhonic tube, from the *Western Home Monthly*, 1925.

Fig. 38.   Sample of Northern Electric tube cartons

Figure 38. The DX-235 is shown as the "Nor-Phonic" tube. Peanut tube cartons are shown from both the NE and the Victor/Northern Electric eras of production. A NE 12SK7GT/G carton is shown from more recent times. This tube, produced in the 1940's, is marked "product of" NE. It is not clear whether it was a product of their manufacturing or of their warehousing operations, but it looks much like an R.C.A tube. Canadian General Electric and Canadian Westinghouse were the predominant radio tube manufacturers by this time, and NE radio tubes were almost unheard of.

Speakers

NE headphones and speakers were normally very close in design to those sold by Western Electric. Headphones came in models R-6A, R-6B (both 2500 ohms), R-6C, R-6D, R-6E (all 3000 ohms), R-7, R-8 (both 1250 ohms), R-10 and R-10A (both 3000 ohms). Not being a student of subtle differences in headphone design, I would say that they look very similar to each other. Models R-7 and R-8 are single ear pieces.

The NE R518 Loudspeaking Receiver was an exact copy of the Western Electric 518W horn, except for the labels. The advertising stated that the NE loudspeakers were designed by the Western Electric Company, New York. The NE R500 speaker was described as a Library Phone, and was similar in appearance to the Western Electric Shawphone. However the R500 stood 19.5 inches high whereas the Shawphone was 14 inches. A third NE horn of the period was the R6900 Loud Speaker, advertised with the R-24 receiver. It came with a rectangular metal base which held the driver, and with a horn above. The arrangement appears like a Hart and Hegeman After Dinner Speaker, but is different in detail and is finished in medium brown. Where the

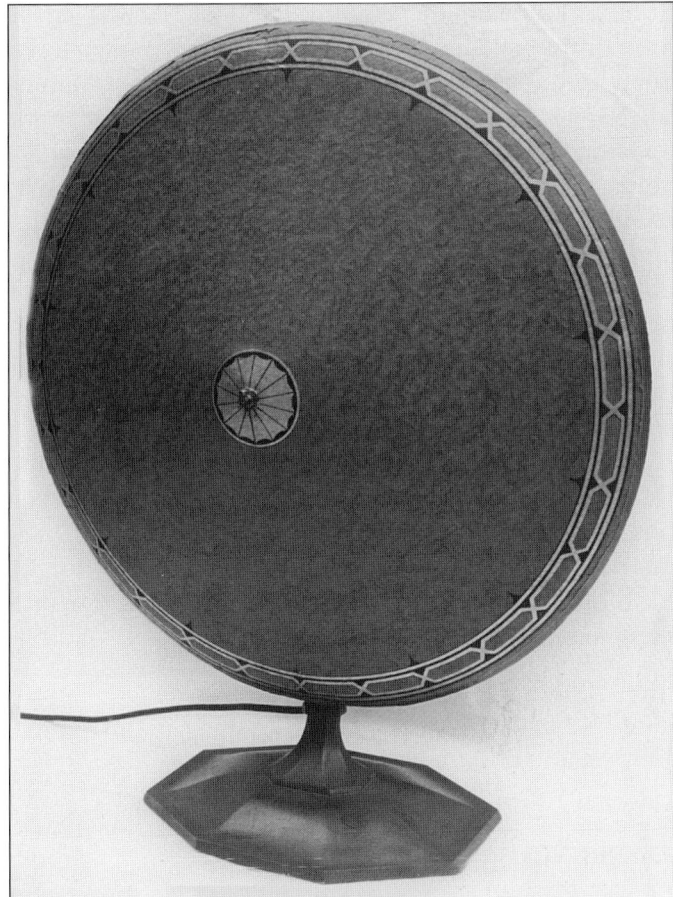

Fig. 39.   Northern Electric R540 cone speaker

Hart and Hegeman base is 5" x 7", the NE base is 4.25" x 5.75". No Western Electric equivalent has been found.

The NE paper cone speaker, model R540, was equivalent in design to the Western Electric model 540AW. It is shown in Figure 39 and is more decorative than its Western Electric cousin, with its pattern created in gilt paint.

Summary and Conclusion

The NE company saw a revenue opportunity with the coming of broadcast radio, and jumped in vigorously at the first. Still it amounted to a small part of the NE manufacturing business. Their early designs were a strong reflection of Western Electric engineering practices of the time. Some of the equipment was designed with direct help from Western Electric and some was evidently designed by NE engineers, one of whom was Cartmel.

In the late 1920's the telephone industry style of engineering was less appropriate for the domestic market. The high reliability caused excessive costs. Cabinet designs in the marketplace were beginning to be more appealing, and NE addressed this change for a while by teaming up with the Victor Talking Machine

Company of Canada Limited. By 1927 NE was again looking for a change in direction. There may have been some pressure from Victor in the U.S. to sever the Victor/Northern Electric collaboration, but there appears to be no record of this. Demand for a.c. operated receivers was growing. Up until 1926, NE receivers relied entirely on Bell System engineering, and Western Electric did not develop their first a.c. operated tubes until 1929. NE ceased manufacture of broadcast receivers in 1926, and then reverted to its role as seller of third party radios in 1931. American Bosch was one such supplier, and their radios were sold in Canada with escutcheons marked NE.

AT&T was prevented by their agreement with R.C.A. from competing in the broadcast receiver business. Perhaps the early broadcast receiver production of NE gives us some idea of how they might have proceeded in the early years, if the Radio Group agreement had been negotiated differently.

Acknowledgments

The author wishes to thank John Radcliffe (curator, retired), Lise Noel and Lorraine Croxen of the Bell Canada Historical Collection, Montreal for their help on my several visits. John not only provided much information but also proofread a draft of this manuscript. Their Collection includes the Northern Electric Collection. I also wish to thank Fred Hammond and the Hammond Museum of Radio, Guelph, Ontario. Collectors Robert MacIntyre of Winnipeg, Lloyd Swackhammer of Alma, Ontario, and Lew Bodkin of Toronto also supplied useful historical information.

At the time of first writing in 2000, all of the items pictured in this article were in the collection of the author. At the present time, the following figures represent artefacts in the collection of the Canada Science and Technology Museum: Figs. 1, 4, 7, 12, 20, 25, and 27.

REFERENCES

1. Lewis, T.S.W. *Empire of the Air: The Men Who Made Radio*. New York, NY: HarperCollins publishers, 1991.
2. Personal communication, John Radcliffe, Curator (retired), Bell Canada Historical Collection, Montreal, Quebec, Canada.
3. Grosvenor, E.S. and Wesson, M. *Alexander Graham Bell: The life and times of the man who invented the telephone.* (Chapter 3). New York, NY: Harry N. Abrams, Inc., 1997.
4. McFarlane, L.B. Preamble to a reply to the A.T. & T. Historical Librarian on the history of the Northern Electric Company Ltd. 1935. Bell Canada Historical Collection, Montreal.
5. Powers of the Northern Electric and Manufacturing Company, Limited, as contained in Letters Patent, 13th December, 1895. Bell Canada Historical Collection, Montreal.
6. McFarlane, L.B. Northern Electric and Manufacturing Company, Document No. 24095, 1938. Bell Canada Historical Collection, Montreal.
7. Canadian Interests Pool Patents for Good of Radio. *The Toronto Star*, August 14, 1923.
8. What Northern Electric means to Bell Telephone. *The Financial Times*, Toronto, December 12, 1923.
9. *Electrical Supplies General Catalogue*, No. 5. Northern Electric Company Limited, 1925.
10. Williams, R.O. The Atwater Kent radios. Chapter 3. *The AWA Review*, 1999; 12, 27-57.
11. Douglas, A. *Radio Manufacturers of the 1920's*. Vol. 3. Vestal, NY: The Vestal Press, Ltd. 1991.
12. Cartmel, W.B. What is a super heterodyne receiver? *Radio News of Canada*, 1925; 3(7), 14-16 & 56-57.
13. Newman, P.C. Northern Telecom: Past present future. Mississauga, ON: Northern Telecom Limited, 1995.
14. Website of the Berliner Museum, Montreal QC. www.contact.net/berliner/ Dec. 16, 1999.
15. Magers, B. *75 Years of Western Electric Tube Manufacturing*. Tempe, AZ: Antique Electric Supply, 1992.
16. Tyne, G.F.J. *Saga of the Vacuum Tube*. Indianapolis, IN: Howard W. Sams & Co., Inc., 1977.
17. Murray, R.P. "The Voice of the Prairie": A Brief History of W.W. Grant (1892-1968). *The Old Timer's Bulletin*, Antique Wireless Assn. 1992; 33(3), 16-19.
18. Allen, B. The Mercury Super Ten. *The Old Timer's Bulletin*, Antique Wireless Assn. 1985: 26(3), 18-19.
19. Murray, B. Eaton Single Circuit Set, 1923. *Antique Radio Classified*, 1994: 11(8), 4-5.
20. Vermeulen, D.J. The beginnings of vacuum tube radio at Western Electric. *The AWA Review*. 1998: 11, 104-149.
21. Cartmel, W.B. The peanut tube and how to get the most out of it. *Radio News of Canada*, 1924, 3(1), 24.
22. Stokes, J.W. *70 years of radio tubes and valves*. Vestal, NY: The Vestal Press, 1982.

Radios of the Canadian Independent Telephone Co., Limited

Robert P. Murray and Herb K. Frederickson

In Canada at the beginning of the 20th century, the Bell Telephone Company was expanding rapidly, but was concentrating first on the more heavily populated areas. To fill the need in rural areas, small independent companies were started, often by local doctors. By 1921 there were around 680 non-Bell companies in the Province of Ontario alone.

The Canadian Independent Telephone Co. Limited, (CITCo), founded by William A. Woods, a Hamilton businessman, was incorporated to manufacture and supply equipment to these small telephone operations. An example of a company ad, which appeared in rural publications is shown in Figure 1.

Originally located at 18-20 Duncan Street in Toronto, the company later moved to larger premises on Adelaide Street, also in Toronto. Occupying the same premises was the Canadian Machine Telephone Co., a manufacturer of automatic telephone switchboards and dial telephones. The two companies shared both facilities and employees, and in 1920, formed a subsidiary company, the Canadian Radio Corporation, Ltd. This latter company acquired licenses and patent rights to manufacture radio equipment, including Canadian rights to the patents of Lee de Forest. It did not, as far as we know, manufacture anything.

CITCo owned and operated its own radio station in Toronto, a marketing strategy that was common at the time. It was listed in Canadian radio magazines as station CKCE, operating on 450 meters. This station

Fig. 1. Rural magazine advertisement for Canadian Independent Telephone Co. telephone equipment.

listing appeared from July 1922 until May 1924 although the hours of operation were not recorded for posterity.

The Radio Sets

The Canadian Independent Telephone Co. began manufacturing a line of de Forest radio receivers in 1923, which were sold under the CITCo label. The Everyman crystal set was a close copy of the de Forest set, including the instruction sheet glued to the inside of the lid, as shown in Figure 2.

The company also produced a one-tube regenerative set, the SC-10, and a matching two-tube amplifier, the model S, as shown in Figure 3. The detector and the amplifier are housed in two sturdy oak boxes. The detector unit is 8" x 11" x 8" deep (excluding the bottom molding); the amplifier is 7" x 11" x 8" deep.

In the SC-10 detector, the tuning condenser is located in series between the antenna and the antenna coils. There is a "loading" switch that adds another coil to increase the band coverage. The phone jack is located in the B- line—an unusual location.

In 1923, these sets were advertised in the Canadian radio magazine Radio as the receiver used by British Prime Minister Lloyd George when he made his way across Canada by train. Royalist sentiment was popular in Canada at the time.

The pinnacle of this series of sets influenced by de Forest was the "Universal Receiving Set", a three-tube set with the tubes and coils protruding from the vertical front panel, as shown in Figure 4. The dimensions are an impressive 12" x 18" x 10" deep, with coils and three 201-A tubes mounted on the front panel.

This was also a regenerative set, supplied in my example with Philips type 201-A tubes, made in Holland for the American market. The interior of the set has Fahnestock clips for the B battery connections, and a wooden battery shelf slid into grooves in the base. The Fahnestock clips were labeled with paper tags inscribed by hand in ink, and hanging on loops of fine string. The A battery terminals are accessible on the front panel.

In addition to receivers, in 1923-24, the Canadian Independent Telephone Co. advertised a horn speaker, as shown in Figure 5. It appears to be conventional in design, although it is rarer now than the receivers. The company also made its own headphones, shown in Figure 2, which are similar in appearance to Brandes phones, but are marked with both resistance (R = 2000 ohms) and impedance (Z = 12500 ohms).

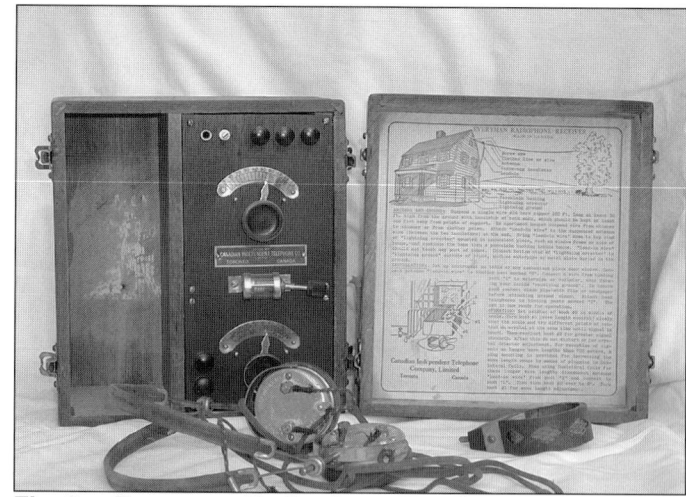

Fig. 2. Everyman Radiophone Receiver

The Last Days

The Canadian Independent Telephone Co. group of companies was experiencing financial difficulties in 1923, and the Canadian Machine Telephone Company filed for bankruptcy on December 28, 1923. Its rural telephone systems were placed under receivership and were eventually taken over by the Canadian Bell system in 1925. On February 2, 1924, a trustee was appointed over the Canadian Independent Telephone Co.

The last advertisements for CITCo products appeared early in 1924, suggesting that the company lasted as a manufacturer of broadcast receivers for about a year. It appears that the three receivers described in this article were the only models made. The Everyman Crystal Set was advertised by Canada West Electric Limited in Regina, Saskatchewan (Exclusive Distributors) for $26.00.

After CITCo went out of business, the Everyman Crystal set was advertised in the Fall and Winter 1925-1926 Eaton's radio catalog for $2.45, and again in the Spring and Summer 1926 catalog. The last listing by Eaton's for the crystal set was in its Fall 1926 catalog and Spring and Summer 1927 catalogs at $1.95. However, on the cover of Eaton's radio catalog for 1930, there is a sketch of a Universal Receiving Set with three Myers RAC-3 tubes mounted on the panel.

Since the Canadian Independent Telephone Co., was one of the first manufacturers of domestic radio receivers in Canada, it attracted the attention of young men seeking a career in radio work. One of these was Ted Rogers, who visited the Westinghouse research laboratory at East Pittsburgh in 1923 on behalf of the Canadian Independent Telephone Co. There he met Frederick S. McCullough and Rogers soon acquired the Canadian rights to the

Fig. 3. SC-10 receiver and Type S amplifier

Fig. 4. Universal Receiving Set

McCullough tube. He also patented some modifications and improvements to it in Canada.

Eventually the Rogers family acquired the Canadian Independent Telephone Company's assets. These included the Canadian Radio Corporation trade name and the de Forest and Armstrong patent rights. The Rogers family went on to build a financially solid communications business that is still in family hands today.

Photo Credit

The artefacts in this article are now in the Canada Science and Technology Museum.

REFERENCE

Chaplin, Maurice. *Rogers Batteryless Receiving Sets, a Brief History*. Manuscript of a presentation at the annual conference of the Antique Wireless Association, Canandaigua, NY, October, 1983. Published privately by the author in 1984. Now Chapter 4 in this volume.

THE NEW
CITCO LOUD SPEAKER

A product of our own laboratory that sets a new standard of quality in Radio reproduction. Designed to faithfully reproduce Radio reception in natural tones without mechanical vibration or distortion of any kind.

The CITCO LOUD SPEAKER is pleasing in appearance, nicely proportioned and elegantly finished.

It requires no auxiliary battery.

When great volume is required power amplification may be used.

An optimum value of tone and volume may be obtained by an adjusting lever, conveniently located on the side.

Special heavy parts do away with mechanical vibrations.

The specially designed horn has a steel stem, with a brass elbow and a 14-inch aluminum bell. This combination of materials produces a tone that has depth and fulness and is not "tinny."

It is easy to attach the CITCO LOUD SPEAKER to any receiving set, as it is equipped with a four foot, high insulated silk covered cord.

Means are provided whereby the wear incident in all adjustable loud speakers may be compensated for, that the device may retain its original efficiency. This is an unique feature and is found only in the CITCO.

PRICE $35.00
Equipped with Cord

Write us to-day for further particulars of this wonderful reproducer. You want the best and we now offer it to you.

CANADIAN INDEPENDENT TELEPHONE CO., LIMITED
TORONTO, - CANADA
Manufacturers of Telephone and Radio Equipment
WESTERN DISTRIBUTORS:
Radio Corporation of Winnipeg, Limited, Winnipeg, Man. Radio Corporation of Reg.na, Limited, Regina, Sask.
Radio Corporation of Vancouver, Limited, Vancouver, B.C. Radio Corporation of Calgary, Limited, Calgary, Alta.

Fig. 5. Advertisement for CITCO loud speaker, in *Radio*, June 1923

This Name Plate on a Radio Set is all ~ Important

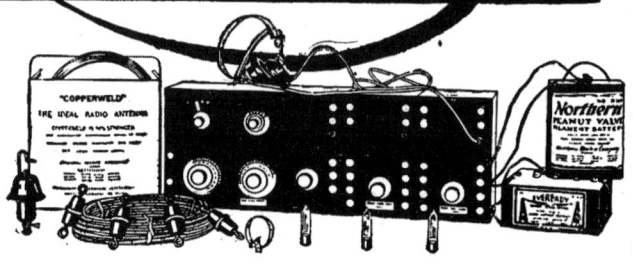

A consistent advertiser stands behind his goods with a guarantee of their sound worth. His name plate is his word, his word his bond. Products which do not live up to their name plates cannot stand the test of consistent advertising.

For many years the Northern Electric Company has spread its goods across Canada through the medium of sound advertising. Its name plate has been a guarantee of perfection.

"Whats in a name!" was never better demonstrated than in a Radio receiving set bearing a trade mark of reliability and sound composition. Years of consistent endeavour in every form of electrical enterprise makes it possible to offer the public a Radio set bearing the name plate of the Northern Electric Company.

Manufactured in Canada by

Northern Electric Company
LIMITED

| Montreal | Quebec | Toronto | | Windsor | Regina | Edmonton |
| Halifax | Ottawa | Hamilton | London | Winnipeg | Calgary | Vancouver |

For sale by Branch Houses and Reputable Dealers.

~ 4 ~

Rogers Radio Ltd.

Ted Rogers, a very capable young radio engineer, visited the Westinghouse Company in East Pittsburgh while on Canadian Independent Telephone Company business. He was to make that trip highly productive. He subsequently secured the Canadian rights to the McCullough AC tube, patented some improvements to it, and promptly took electric light-socket operated sets to market in Canada. These AC operated sets were cheaper to run than their battery precursors, and circumvented the hazard of getting battery acid on the family living room carpet, not a trivial advantage in the 1925 radio market. The vision of Ted Rogers and the business acumen of later family members have ensured a continuing family business in various aspects of broadcasting and communications.

"Just Plug In—Then Tune In"
The First Commercial Light-Socket Operated Radio Receivers from Rogers Radio Ltd., Toronto, Canada.

Copyright © 1984 by Maurice Chaplin
Reprinted by permission.

The early nineteen-twenties saw a major domestic upheaval in most industrialized countries as radio entered the home and changed the lifestyles of thousands. By the end of 1924, however, enthusiasm was waning partly due to the expense and bother incurred in satisfying the "box's" appetite for batteries (Radio battery sales reached $450,000 in the U.S. alone for 1924). This was especially irksome for the majority of town and city dwellers who already had an electric supply wired into their homes for lighting and cooking purposes. A ready market existed for radio receivers able to operate directly from the light socket and many companies and individuals hoped to satisfy that demand.

In the United States the Radiola Model 17 marketed by RCA in 1927 is generally considered to be that country's first all-electric receiver although pioneer radio engineer B.F. Meissner has written that Garod's model EA, incorporating his designs, deserves that distinction having been introduced publicly in May, 1926. The first European socket-powered receiver to gain wide acceptance was probably the Philips 2511 on sale in Holland in 1927. In Canada it is believed that one of their own, E.S. "Ted" Rogers, deserves the credit for introducing the world's first commercial batteryless receiver and building a major electronic communication business from that success.

Rogers' Canadian biographers suggest that his accomplishments should rank him alongside such eminent scientists and inventors as Bell, Armstrong, de Forest and Marconi. However, noted vacuum-tube authority Gerald Tyne in his definitive *Saga of the Vacuum Tube* wrote "... published information pertaining to Rogers' dates of manufacture and establishment of companies is limited and conflicting ... the brochure "Edward Samuel Rogers Collection" is a disservice to Rogers because of inaccuracies. His contributions to the industry and Canadian radio history need no embellishment".

That last sentence gives the spirit and intent of this paper.

Edward Samuel Rogers (1900-1939)

Edward "Ted" Rogers was born on June 21, 1900 to Albert S. and Mary (Ellsworth) Rogers of Hazleton Avenue, Toronto. Both Albert and his brother Joseph were executives with the Queen City Oil Company founded by their father some years earlier. The family moved three times during Ted's early childhood before settling in 1907 in a substantial brick home at 49 Nanton Avenue.

Fig. 1. Edward "Ted" Rogers from *Radio News of Canada*, December, 1925

Unlike his father who had been educated at Pickering College, a Quaker college in which his family had played a prominent financial role, Ted attended Toronto's public schools. A precocious child, Ted developed a passionate interest in the still comparatively new field of wireless telegraphy. He learned the Morse code so that he could decipher the messages received with his progressively more complicated home-made equipment. The Rogers family were Quakers with well defined ideas on what was and was not acceptable behavior and being able to withdraw into the private world of radio communication was probably a welcome relief for a high-spirited child. In April, 1912, during a Spring holiday at the family's cottage on Georgian Bay, Ted set up a receiver and picked up the messages giving the names of the survivors of the Titanic disaster as they were relayed by David Sarnoff. (Considering that Sarnoff may not have relayed them,[1] the source for this bit of information would have to be regarded as dubious - Ed.) Later, in 1914, Ted's receiving station, this time at Nanton Avenue, gained some prominence, when he was able to tell of the War Declaration in Europe before the newspapers distributed the story.

Ted was enrolled in the University of Toronto's Electrical Engineering program but withdrew at the end of his second year probably discouraged by the small part radio communication played in the curriculum. During 1920 he enrolled in the University's School of Practical Science and he graduated in 1922.

As soon as wartime restrictions were lifted in 1919 Ted became the licensed owner-operator of amateur radio station 3BP with a 500 watt spark transmitter at Nanton Avenue. Later he was able, because of his family's connections, to relocate the station to an apartment he made for himself in one wing of the still empty Pickering College (the college had been used as a military hospital during World War I), at Newmarket. From here his signals were received in both the Atlantic and Pacific regions (considered a remarkable achievement at the time). In 1921 he transmitted as part of the December Trans-Atlantic Test and became the first Canadian to span the ocean when his signal was received by Paul Godley at Androssen, Scotland. Ted was able to acquire a de Forest "oscillion" radio tube and was soon transmitting C.W. in place of "spark".

After graduating in 1922 Ted joined the staff at the Canadian Independent Telephone Co. Ltd., as a radio engineer.

The Canadian Independent Telephone Co. Ltd.

At the turn of this century, the Bell Telephone system was expanding as fast as technology and finances permitted but, naturally, it attempted to service the more highly populated areas first, where the capital cost per phone was less and the anticipated revenue greater. In the rural districts Bell could only tell potential subscribers that it might be years before a service was available to them. Many of the smaller communities, therefore, organized telephone systems of their own, with the initiative often being taken by the local doctor. By 1921, some 680 non-Bell companies were operating in Ontario.

The Canadian Independent Telephone Co. Ltd., was organized by a Hamilton businessman, William A. Woods, to manufacture and supply equipment to these small telephone-system operators. They ran advertisements in the rural press—the *Farmer's Advocate* or *The Country Gentleman*, and usually had an exhibit at the Canadian National Exhibition held annually in Toronto.

Initially located in Toronto at 20 Duncan Street the

Fig. 2. Magazine ad for the Canadian Independent Telephone Company

Company moved to larger premises on Adelaide Street. Sharing this address was a related business, The Canadian Machine Telephone Co., which had originated at Brantford, Ontario, in 1900 as a manufacturer of automatic telephone switchboards and dial "machine" telephones invented by Romaine Callender. Eventually two company employees, the Lorimer brothers, became the owners. In addition to manufacturing and selling equipment to customers across Canada, in France and in Italy, the company operated a number of small telephone systems at various rural Ontario locations. The division between these two companies blurred as facilities were shared and individuals assumed positions of responsibility on both Boards.

With an eye to the future, a spin-off company—The Canadian Radio Corporation, Ltd., was formed in 1920. It began to acquire licenses and patent rights for manufacturing radio equipment. Amongst those held were the Canadian rights to the inventions of Dr. Lee de Forest and Edwin Armstrong.

Early in the 1920's all three companies had their registered offices and showrooms at 212-216 King St., Toronto—a building which also housed offices and showrooms of the Canadian General Electric Co. Manufacturing space was also leased from CGE at their Ward and Wallace Avenue factory. An experimental 2,000 watt radio transmitter, CKCE, was also operated by the Canadian Independent Telephone Co., from the latter address.

The Canadian Independent Telephone Co. was the first, and possibly the only, manufacturer of commercial radio equipment in Toronto around 1922, and for young men seeking a career in this field, it was the place to work. The company built CITCo brand horn-loudspeakers and assembled a range of de Forest radio receivers which were then sold under the Canadian Independent Telephone Co.'s label. (A more detailed description of CITCo radios has been published.[2] - Ed.)

Toronto's first commercial radio station, CFCA, owned by the Toronto Star newspaper, started

Fig. 3. Ad for radio products of CITCo, *Radio News of Canada*, January, 1924, p. 51

broadcasting on March 28, 1922. The equipment for the station had been installed by the Canadian Independent Telephone Co., and one of Ted Rogers' jobs was to work with another engineer, Dr. Charles Culver, on improving the transmitter's performance. The Heising modulation system was adopted with a Colpitts oscillator circuit. CFCA used a Ford Model T panel truck as a radio car for publicity purposes. Ted was the operator at many of the shows and exhibitions where the station quickly (and loudly) made its presence known.

Ted Rogers was also involved with solving the problems associated with the operation of radio equipment from a home-lighting power source. He was, no doubt, keeping abreast of developments made elsewhere through reading the articles in the popular radio magazines as well as the more technical reports in the *Proceedings of the Institute of Radio Engineers* available through Dr. Culver.

The Batteryless Radio — Early Days

As the novelty of radio in the home diminished, owners became dissatisfied and demanded the better sound quality which led to improved loudspeakers. These loudspeakers had to be driven by more powerful amplifiers which then required still larger, more cumbersome and more expensive batteries for their operation. Most town and city dwellers already had access to an alternating current electrical supply for home lighting and power purposes and it seemed quite logical and simple to the average owner that the radio receiver, an electrical device, should also be powered from this source. The market was ready for the batteryless receiver— why the delay?

In order to answer this question, we need to consider a bit of technical radio theory. A loudspeaker converts changes in electrical current into changes in air pressure. If the current changes at an audible rate so will the air pressure which, when heard, becomes sound. The loudness level will be proportional to the current's intensity.

In a radio system the loudspeaker's current changes should be caused only by the program content of the received signal. The early receivers achieved this by having the radio signal cause changes in the otherwise steady current produced by batteries of electrochemical cells. Three of these battery systems were generally required: a low-voltage, high-capacity type known as the "A" battery, which supplied the energy to raise the temperature of the filament used in each vacuum tube to obtain a supply of liberated electrons within the tube; a "B" battery, one of smaller capacity but much higher voltage which provided the steady current to be varied by the signal; and a low-voltage, low-capacity, "C" battery needed to provide the correct voltage/current relationships for minimum distortion.

The first steps taken to electrify the radio involved replacing the "B" supply. "B"-battery eliminators were developed. These units included a transformer to both provide voltage step-up and ac supply isolation, a rectifier, filter, and a divider to obtain the different voltage levels required in the receiver—a function previously taken care of by using one or more of the taps provided on the "B" battery.

"B"-type batteries were generally made up from a number of carbon-zinc primary cells but lead-acid cell batteries were also available, the higher initial cost was offset by greatly extended life as the battery was capable of being recharged.[3]

When a radio receiver was purchased in the early 1920s the vacuum tubes had to be bought as a separate package. The customer had to choose between those that used dry-cells for their filament supply (requiring frequent battery replacement), or those that would operate directly from the larger, heavier, more expensive, automobile-type storage batteries. This type of battery could, of course, be recharged when necessary using the services of the local garage, a home-charging unit, or by simply exchanging it for a period of time with the one used in the family car.

To eliminate the "A"-battery with a unit similar to the one replacing the "B"-battery was feasible but not practical since the high "A"-supply current demand could only be met by using large, heavy, expensive and cumbersome components.[4] The ultimate receiver of the day appeared to be one which contained within its battery compartments a "B"-eliminator and a small storage-type "A"-battery with a charging unit which would automatically cut-in when the receiver was not in use.[5] The storage battery, however, with its corrosive acid and volatile fumes was not a pleasant home companion.

An alternative to having the receiver use dry-cell type tubes was for the interior wiring to be changed so that the tube filaments were series-connected. This resulted in reduced "A"-supply current demand making operation from a "B"-type eliminator feasible. A number of difficulties arose, however, with series-filament operation so that this was not the simple solution it may, at first glance, have appeared to be.

Instead of a modified battery-receiver, what was needed was a radio designed specifically for AC line operation. The main problem source lay in the vacuum tubes of the day so the re-think began there.

Vacuum Tubes Using AC as a Filament Supply

If a conventional triode tube such as, for example, a 201-A is operated with an AC supply for its filament source the tube's plate current will have a large line-frequency ripple current even though the tube's "B" and "C" potentials are obtained from a steady DC supply (see Figure 4).

Three factors combine to produce this effect.

They are:
1. A voltage gradient exists along the length of the filament and if one end of the filament is common to the "B" and "C" supplies all these potentials will vary at the supply frequency.
2. The filaments' I^2R heating effect is reaching its maximum and minimum values at twice the supply frequency. Electron emission, a function of temperature, will vary at the same frequency.
3. The electron flow within the tube will be affected by the varying magnetic field surrounding the filament.

The effect of the first factor is minimized by returning the tube's input and output signal circuits to the electrical centre of the filament either with a tapped filament or to an artificial tap obtained from a tapped resistor—known as a humdinger—connected across the filament circuit. A variable resistor gave more precise results (see Figure 5).[6]

The cyclic heating factor is reduced by having an emitter with "thermal inertia" (one that will resist rapid temperature changes). The first solution was to use a thicker wire filament, its greater mass producing the desired effect. The lower resistance of the thicker wire needed a larger current intensity to produce the heat required but the voltage was smaller helping still further in offsetting the effects of the first factor mentioned above. [7]

The hum caused by the filament current's chang-

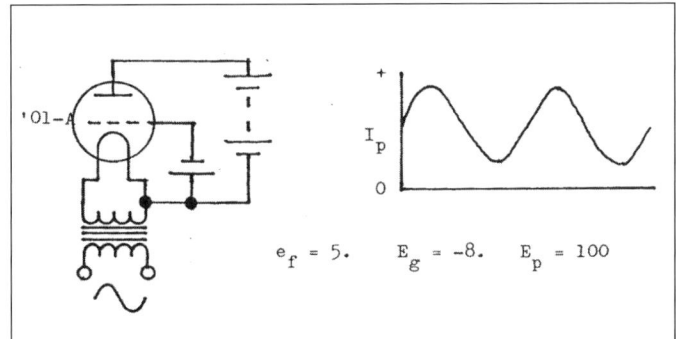

Fig. 4. Illustration of AC ripple in a DC tube.

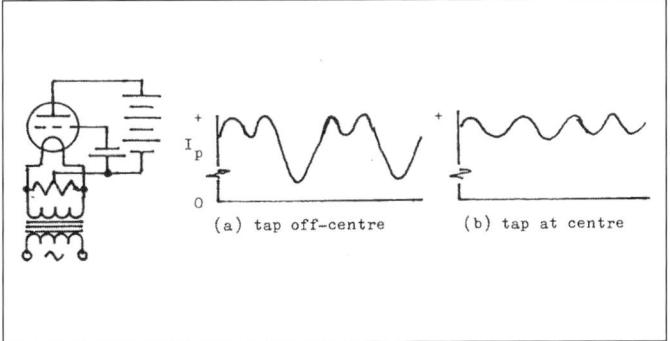

Fig. 5. The effect of a tapped resistor across the filament.

ing magnetic field could be reduced by cancellation if a hairpin-shaped or bifilar-wound filament was used instead of the straight type. Although the increased rate of current flow accompanying the thicker filament should increase the magnetic problems a careful choice of filament voltage-current values could produce good hum reduction.

The larger current requirement does, however, add to the overall receiver design problems when a large number of parallel-connected-filament tubes are in use. The strong fields produced must be balanced out using magnetic shielding techniques to reduce inductive hum pickup.

The vacuum tube so far described is suitable for use as an amplifier of either radio or audio frequencies but cannot be used in the detecting stage of a radio receiver. For detection the tube must operate in a non-linear region where self-cancellation features do not apply. A number of radio receiver designs, therefore, used the AC heated tube in the amplifying stages and used either a crystal detector or a dry-cell type vacuum tube with its filament energized with DC obtained from the "B" supply.[8]

The Indirectly Heated Vacuum Tube

Another way to avoid the voltage gradient along the emitting surface was to have an emitter heated from some separate source. The emitter might be of the Wehnelt type[9] with a lime coating on a platinum tube. Within this cylinder was a filament wire which, when connected across a suitable supply, acted simply as a heater for the cylinder. A vacuum tube using this principle was patented by H.J. Round of the Marconi Company.[10] After its disclosure, John Scott Taggart, a British radio authority, is quoted as saying "... the Marconi device, though experimentally successful, has no commercial value ...". The filament, incidentally, was a straight wire running axially within the cathode cylinder and had to be carefully positioned so that the two did not touch. The cathode's temperature had to be raised by radiation from the heater and time delays up to 100 seconds were not uncommon from the time of first applying heater power to obtaining adequate electron flow.

A British engineer, Alexander McLean Nicolson, working in the United States for Western Electric, had been involved with this type of equipotential cathode since 1913 and applied for a patent two years later.[11] His cathode consisted of a cylinder of quartz or similar refractory material with a platinum layer deposited and baked onto its surface. An oxide coating was then applied over the platinum and the heating wire passed down through the center of the quartz tube. The indirectly-heated tube was effectively covered by the 23 claims of this invention.

On March 23, 1918, Hubert W. Freeman applied for a patent on an indirectly-heated vacuum tube.[12] Assigned to Freeman's employer, Westinghouse, the idea was placed on the back burner as the company was then fully occupied with thoriated-tungsten filament development.

Samual S. Torrisi of Philadelphia also designed an indirectly-heated tube with the grid, cathode and plate electrodes supported within the tube in the conventional manner but with the heater insertable from the top of the tube so that it could be readily replaced when necessary—the heater's connections at the top of the tube gave it a double-ended form of construction.[13]

Production samples of Freeman's tube were made by Westinghouse at their East Pittsburgh plant during 1922 and several experimental circuits were devised. The results were disclosed in an article Freeman wrote for publication in the December 1922 issue of *The Electric Journal*. Headed "A Practical Alternating-Current Radio Receiving Tube", the article itemized the reasons why alternating-current was unsuitable as an energy source for radio tubes of the emitting-filament type. It showed how an equipotential cathode eliminated the problems when used with a correctly-shaped heater, and gave full details on the experimental tubes. Also given were the diagrams for receiving circuits used in the tests, and cautionary notes on the importance of keeping all leads carrying AC away from the signal circuits, and the increased tendency for hum pickup when two or more transformer coupled audio-frequency stages were used. Freeman recommended retaining batteries for the "B" supply as "... the supply requires delicate adjustment for satisfactory operation ... and batteries are simple and comparatively inexpensive ...".

Freeman and a co-worker, Wallace Wade, filed for a US patent in 1922,[14] a British patent in 1923[15] and a Canadian patent in 1925. Although the application shows a single-ended tube with the cathode connection brought out to a fifth pin, the first Westinghouse production samples were double-ended. These were designated type UX 225.[16]

John Morecroft in the 1921 first edition of his *Principles of Radio Communication* mentioned using an equipotential-cathode vacuum tube with AC applied to its heater. Dr. A.W. Hull of General Electric patented a circuit based on this idea in 1922 where an indirectly-heated tube was used as a combined

— 74 —

rectifier and detector-amplifier.[17]

Incidentally the idea of using a double-ended tube with a removable heater was revived by Dr. A.N. Lucian of the University of Pennsylvania in 1926. He also suggested that AC could be used as the heater supply.[18]

The Canadian Independent Telephone Co. was interested in line-operated radio receiver development and in 1923 Ted Rogers visited Westinghouse's research laboratory at East Pittsburgh, Pa. There he saw experimental tubes in various states of development and discussed with their engineers some of the problems being encountered. One of these engineers, Frederick S. McCullough, held patents in his own name on several vacuum tube developments and manufacturing processes.

Frederick S. McCullough

Preparing biographical notes on Frederick S. McCullough is difficult as very little information is available. However, the information we have would suggest that he was clever, energetic, ambitious and, probably, opportunistic.

McCullough lived in Cleveland during World War I and was employed as a radio engineer with the Glen L. Martin Company. He was mainly concerned with developing airborne radio equipment for navigational purposes and he obtained several patents which were assigned to his employer. One of these[19] shows copper coils below the surface of a flying-field runway designed to radiate RF energy as part of a guided-approach landing system. Another shows his awareness of the problems arising from distributed capacity as he attempted to improve RF amplification in airborne radio receivers by having the coupling coils within the vacuum tubes.[20]

The aircraft industry went into a tailspin after the war ended and, in 1920, McCullough went to work as an engineer with the Westinghouse organization at their East Pittsburgh plant. He lived, as did many other Westinghouse people, in the neighboring suburb of Wilkinsburg.[21] He was mainly involved with vacuum-tube research and production technology, bringing additional expertise to Westinghouse as a result of the fact that he had developed the RF induction heating method for liberating electrode impurities some years earlier.[22]

Elected to the IRE as an Associate in 1919 and a Member in 1922, McCullough presented a paper at the New York meeting in September, 1922.[23] Here he spoke on the construction and characteristics of vacuum tubes, using the recently-introduced WD11 as his example. He then discussed the advantages of equipotential cathodes and, finally, he gave details of tubes he had constructed and tested at Westinghouse. Two of these used AC as a cathode heater supply—one across 110 volts the other using 5 volts at 2 amps.

A few days earlier McCullough had applied for a Canadian patent on a Cathode Device and a Circuit which employed a vacuum tube with an equipotential cathode heated inductively from a coil connected to an RF generator. This coil could be within the tube or mounted externally. The tube's 2 plate electrode was segmented to reduce inductive heating within it.[24]

Towards the end of 1922 McCullough had applied for three U.S. vacuum-tube patents on his own behalf, followed in March, 1923 by one for a multiple triode tube which was assigned to his employer, Westinghouse.

Rogers Radio

Many of the experimental tubes Ted Rogers saw at the Westinghouse plant incorporated improvements attributed to McCullough. Convinced that this tube was commercially viable and that the production difficulties could be overcome, Rogers made private arrangements to acquire the Canadian manufacturing and sales rights to McCullough's patents and developments for a sum quoted as between ten and fifty thousand dollars. A number of Canadian patents were applied-for and assigned to Albert S. Rogers, Ted's father. The first few do not appear to be too valuable—some being Canadian rights to inventions already patented in the U.S. and assigned to Glen Martin but they do establish the licensing agreement between McCollough and the Rogers family. Two later patents are for metal envelope tubes with the envelope acting as the plate (shades of Catkin construction), and for methods used to improve glass-to-metal seals.[25]

On November 2, 1923 Ted Rogers applied for a patent on an inter-stage tuned variable-coupling system[26] and assigned it to the Canadian Radio Corporation, the CITCo subsidiary. These companies were experiencing financial problems and the Canadian Machine Telephone Co. filed under the Bankruptcy Act on the 28th of December. Its rural telephone systems were placed under trusteeship and were eventually taken over by the Bell system in 1925. A trustee was appointed over the Canadian Independent Telephone Co. Ltd. on February 2, 1924. An arrangement was made with the creditors and the companies assets, including the Canadian Radio

Corporation tradename and its all-important de Forest and Armstrong patent rights, were acquired by the Rogers family and moved into the former T. Eaton Co. warehouse at 90 Chestnut Street in downtown Toronto.[27] CITCo and its subsidiaries were listed as being at this address for the remainder of the year but little telephone work was done as the new plant was readied for radio tube and receiver production.[28]

The Rogers Radio Company was formed early in 1924 with A. S. Rogers as President, F. S. Rogers as V.P. and Samuel Rogers as Secretary. The registered office was Room 405 at 56 Church St., Toronto, (the private office of Albert Rogers) in the Imperial Oil building. Two limited-liability companies were registered on November 28, 1924. These were Rogers Radio Ltd., with the same executives and offices and Heliotron Tubes Ltd., a manufacturer, importer and exporter of radio tubes.

The McCullough Tube

Canadian patent 256,378 issued to McCullough in April, 1925 is of interest as it shows an early attempt to solve the problems associated with using AC as a filament supply. The tube has an emitting filament wound around a porcelain rod which acts as a heat reservoir. Two versions of filament are shown. One is bifilar wound and the other has one end of the filament brought down within the porcelain rod.

Patent 265,021 shows a device more closely resembling what is now known as the McCullough tube (Figure 6). It shows a long porcelain rod with a coated metal cathode sleeve, a grid and a plate at one end. Within the rod but at the other end is the resistive heating element. The electrical fields from the heater current are thus kept away from the tube's signal electrodes. The heater's lead wires are brought out at the top of the tube envelope and fitted within a cap similar to the double contact base used with automobile lamps. This allowed the tube to be quickly disconnected from its external circuits.

On April 25, 1925 Earl L. Koch of Pittsburgh, applied for a U.S. patent disclosing a tube which was identical in appearance to McCullough's but having the heater down within the cathode sleeve. A second grid located below the signal-grid had a small negative potential applied to it and this, it was claimed, greatly reduced induced hum. This patent was assigned to McCullough.[29]

As part of a May 6, 1925 patent application for a tube de-gassifying method, McCullough shows the heater within the cathode sleeve as in the Koch patent but the auxiliary grid is omitted.[30] This grid does, however, appear in a Canadian application filed May 29, 1925.

The modifications needed in the filament circuits of existing battery operated radio receivers to accommodate the McCullough tube are detailed in one patent application.[31] Numerous other patents deal with tube manufacturing methods, in baking the emitting surface,

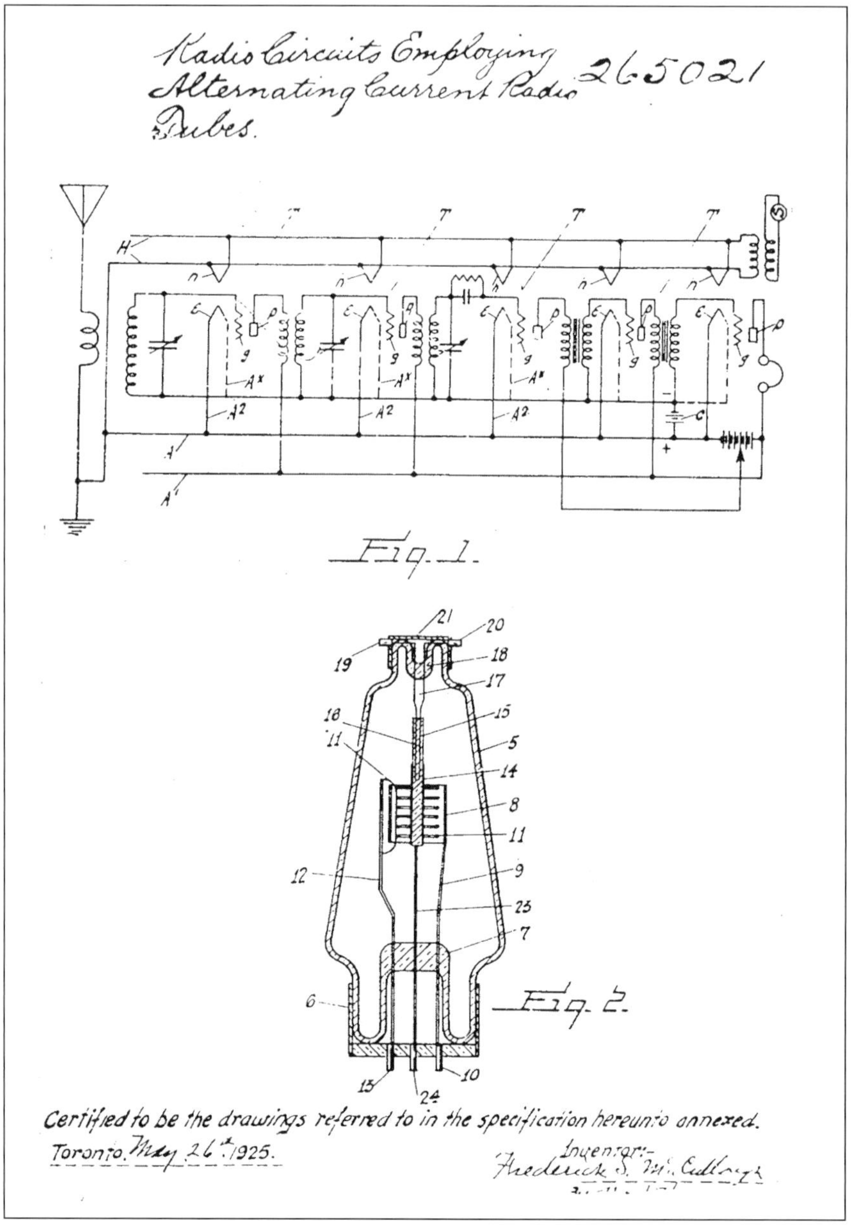

Fig. 6.   Illustration from Canadian patent 265,021 awarded to McCullough

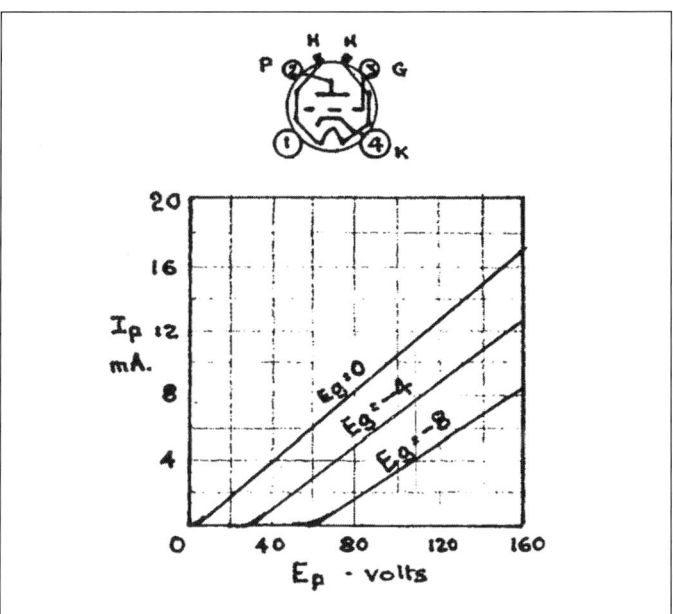

Fig. 7. Characteristics of the McCullough tube

de-gassifying, and temperature compensation to obtain compatibility between tube electrodes and the internal insulators.

As readied for production the McCullough tube had a brass, short-pin Shaw base with the name McCullough and the emblem AC with two lightning flashes on the brass skirt. The top cap was of bakelite with what would have been the two bayonet locking pins now serving as the heater connections. The tube had no type number.

Provisional electrical specifications were:
Eh = 3 V. Ih = 1 A. Ep = 90 V. Ip = 4. mA.
Eg = - 4.5 V. Zp = 9k5 ohms. gm = 870 mho.
amplification factor = 8

McCullough Tube Sales - U.S.A.

The license fees received from the Rogers family enabled McCullough to leave Westinghouse and become self-employed. He moved into a new home, and then formed the F. S. McCullough Company Research Laboratory in premises formerly occupied by the King Radio Manufacturing Company at 521 Penn Avenue W., in Wilkinsburg, Pa.[32]

There were two distinct potential markets for the McCullough tube. They could be incorporated as original equipment into the next season's radio receivers designed specifically for line operation or, they could be used to replace the popular 201-A used in thousands of battery-operated receivers so that the radios could be converted (with a suitable transformer and "B"-eliminator) for AC line operation. The latter promised the fastest financial returns and was the one first addressed with national advertising.

The largest circulation radio magazine, *Radio News* (250,000) had paid scant attention to McCullough in its news or editorial columns so the main advertising was placed with the second largest magazine, *Popular Radio* (95,000). Two separate full-page advertisements appeared in the May, 1925 issue. In one, Radio Foundation Inc. of 25 West Broadway, New York City, announced that they were exclusive agents for the McCullough Sales

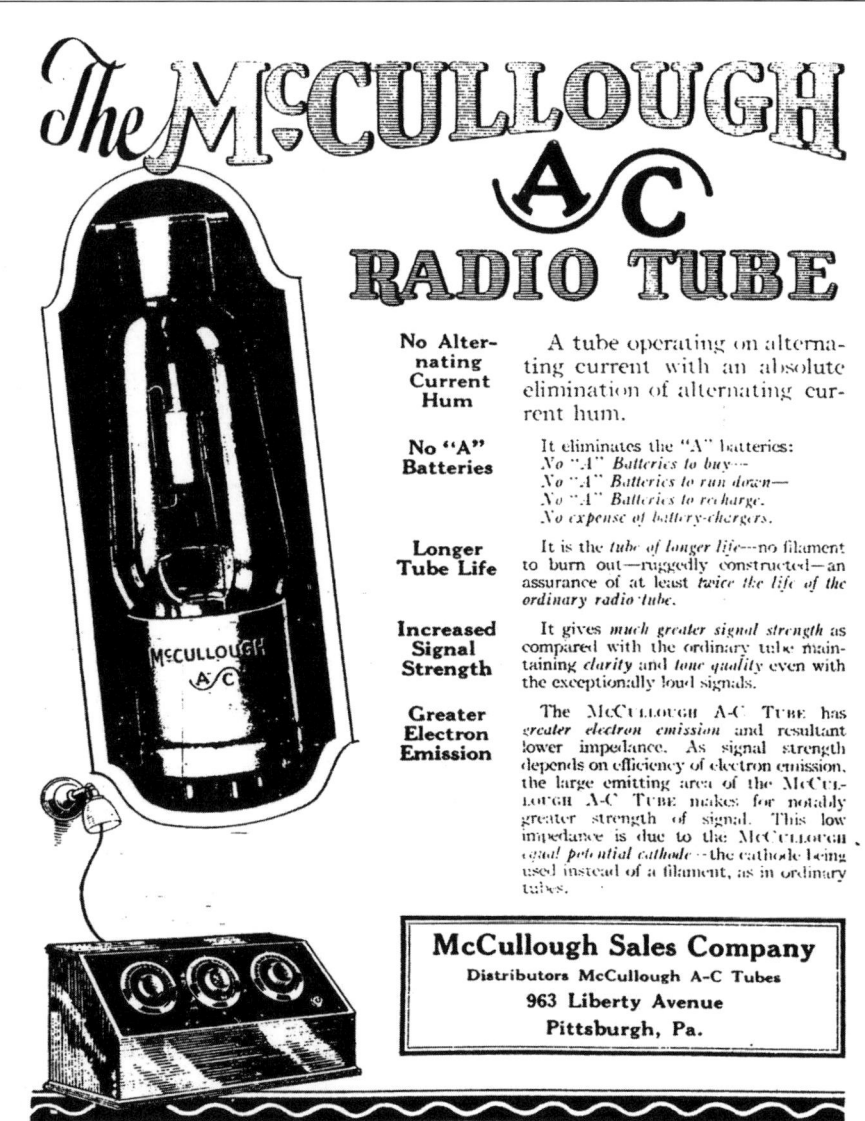

Fig. 8. McCullough AC tube ad from *Popular Radio*, June, 1925

Fig. 9. McCullough/Freshman radio ad from *Popular Radio*, June, 1925

Company and invited jobbers to apply for area franchises. The second advertisement was by the McCullough Sales Company at 963 Liberty Ave., Pittsburgh. This was the address of an established jobber, the Pittsburgh Radio Supply House.

Popular Radio's June, 1925 issue carried a two-page advertisement from McCullough Sales claiming "... an absolute elimination of alternating current hum ...". The tube illustrated now had on its base a sine waveform around the letters AC. The list price was $6. Also in this issue the tube is mentioned in an article on new tube developments and in a reader's query on the Q and A page. Several pages are devoted to plans for constructing a receiver around the tubes.[33] Amongst the advertisements, four suppliers list complete parts kits[34], one manufacturer shows a heater transformer[35], and four manufacturers offer complete receivers (each of these requiring an external "B"-supply).[36]

Two-page advertisements for the tube were placed in the July, 1925 issues of both *Popular Radio* and *Radio Broadcast*. In the same month, *Radio News* had an article on the tube and *Popular Mechanics* published construction plans for a 5-tube receiver designed around the tubes. However, in an article on the Principles of AC Tubes appearing in the August, 1925 issue of *Radio*, E. R. Turner writes "In practice the hum has not been completely eliminated from such a circuit due to induction from the ac leads and capacity between the heater and the cathode." Very little more was said about the tube in 1925 other than mentioning it in another construction article, this time for a 3-tube receiver, in the December issue of *Popular Mechanics*.

The demand for the tube was heavy and McCullough with limited production facilities could not meet delivery dates. Quality control was marginal and customer complaints multiplied. Possibly, in addition, McCullough may have been receiving information on Westinghouse's progress with their indirectly heated tube. For whatever reason, McCullough sold his manufacturing rights to the tube in late 1925. The transfer to the well-established Kellogg concern was announced in January, 1926.[37]

In February, 1926 McCullough invited radio manufacturers to incorporate the tubes in their designs and offered to wire prototypes in a Brooklyn, New York, laboratory.[38] The Cleartone Radio Company took advantage of this and claimed, in September, 1926, to be the first to produce a completely self-contained line-operated receiver with their model 110. Their claim was modified a month later to read " ... the first successful low-priced set to eliminate "A" and "B" batteries".[39]

This quotation from a paper delivered by Julius

Aceves to the Radio Club of America is indicative of the adverse publicity being encountered: "... unfortunately these tubes are not made with any degree of uniformity, and although their life should be theoretically much longer than thoriated tubes, it may be only one month, after which the emission has been reduced to a useless value or the heater may burn out in a few weeks. If the McCullough tubes were properly made, they would unquestionably be the tubes of the future, and radio sets would be designed for them on account not only of the complete elimination of the "A"-battery but of their inherent high amplification with low resistance properties".[40]

The tube now had a UX-type base and was marketed as the McCullough 401. Then, as improvements were made to the heater-cathode insulation, the name Kellogg appeared on the tube. In addition to being used in receivers manufactured by Cleartone and Kellogg the tube was fitted in Bush & Lane, Dayfan, Marti, Mohawk, Shamrock and Sparton models for the 1926-27 season. (Sparton renamed the tube "Cardon"). An output tube type 403 was also made by Kellogg, until they discontinued production in 1929. The AC Neon Co. took over manufacturing with the 401 now listing at $5 and the 403 at $7.50 (more expensive than the now more popular RCA UX-227).[41]

Frederick S. McCullough continued as an independent research engineer and, although he dropped his membership in the IRE, he continued to file patent applications for vacuum tube improvements into the late 1930s. An unconfirmed report had him last active in a radio venture on the West Coast.

Early Rogers Vacuum Tubes

In 1924 Ted Rogers and the fledgling Rogers Radio Ltd. were fully committed to setting up 90 Chestnut Street to manufacture radio tubes and receivers. The first series of tubes produced were diodes to be used as rectifiers in "B"-power supplies. These tubes, later known as type R-100 or RX-100, are structurally similar to a 201-A with the grid removed (Figure 10). The filament was rated at 5 volts and 0.60 amps. A version with a 1.25 amp filament was labeled the RX-200 (Figure 11).

Samples of the McCullough tube were available, jigs were made up, vacuum pumping, flashing, aging and testing equipment was devised and set up and production samples trickled through for evaluation. Some problems were immediately apparent. The warm-up time, for example, was longer than anticipated and flaking of the emitting surface material showed the need for improved bonding. Other difficulties encountered (in some tubes showing up immediately, in others only apparent over a longer term), were control-grid emission, heater to cathode emission, heater-cathode leakage, deterioration of inter-electrode insulation, and premature heater failure.

With the technology then available, it appeared that compromise solutions to these problems were all that could be anticipated (Attempting to decrease warm-up time by increasing the heater power led to increased difficulties with grid emission and early heater failure). Reducing the diameter of the twin-bore ceramic insulating rod introduced problems if the bore was not sufficiently oversize to accommodate the expansion accompanying the heater's temperature change and also added to leakages. Promising results were obtained when the grid and plate electrodes' surface areas were increased to dissipate unwanted heat and when other refractory materials were used as cathode insulators. Ted Rogers worked on solving these problems practically single-handedly. A steady stream of advice was coming from McCullough together with a number of patented ideas. Most when tried, however, introduced still more problems.

Fig. 10. Rogers R-100 rectifier

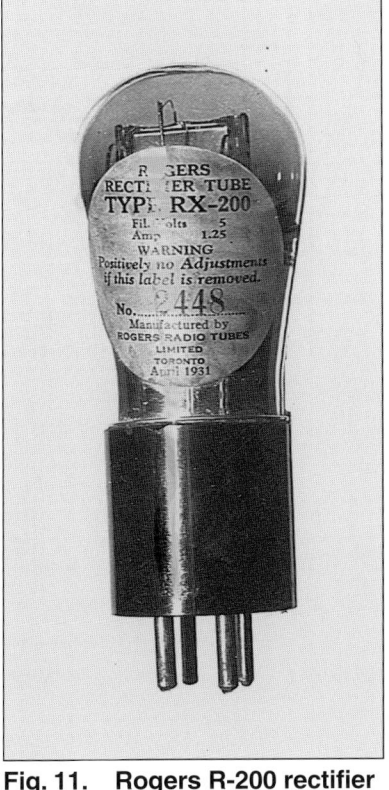

Fig. 11. Rogers R-200 rectifier

The Rogers Patents

Three Canadian patents were applied for by Rogers during 1925. All three were granted. They were: 250,714 (Feb. 21) for a rectifying system (Figure 12). The disclosure shows a transformer-operated full-wave circuit using two thermionic diodes. In addition to the conventional low-pass filter system, RF chokes have been added in each of the "B"-supply leads together with suitable by-pass capacitors. These components were included to prevent modulation hum that could occur as the signal passed via the L.F. choke's distributed capacity to the rectifying diodes. The patent application shows this supply providing "B"-potentials to a two tube (filament type) regenerative receiver which is, incidentally, described as a three-tube in the application.

A two-tube receiver circuit using indirectly-heated tubes from an AC supply is shown in Patent 264,940 (March 18). The tubes act as combined amplifiers and rectifiers and the circuit appears to be very similar to one described by Dr. A. W. Hull in the April, 1923 issue of the *Proceedings of the I.R.E.*

Patent 269,205 for Improvements in Radio Reception and Amplification shows a three-tube receiver circuit diagram and component layout (Figure 13). Indirectly heated tubes with AC applied to their heaters are shown but the "B" and "C" potentials are obtained from batteries. The two audio-frequency coupling transformers are both contained in a sheet-metal shielding can and are located within the cabinet some distance from the tubes' AC heater wiring to minimize inductive pick-up. The position of the line-operated heater transformer is not shown.

Standard Radio Manfacturing, Corporation Ltd.

Progress was being made with improving the first indirectly-heated tubes and many of acceptable quality were produced. These were imprinted as "Canadian McCullough AC Radio Tubes. Patented 1923-25. Other Patents Pending". This was later revised to read "Can. McCullough Radio AC Tube. Made in Canada by R.R.L.

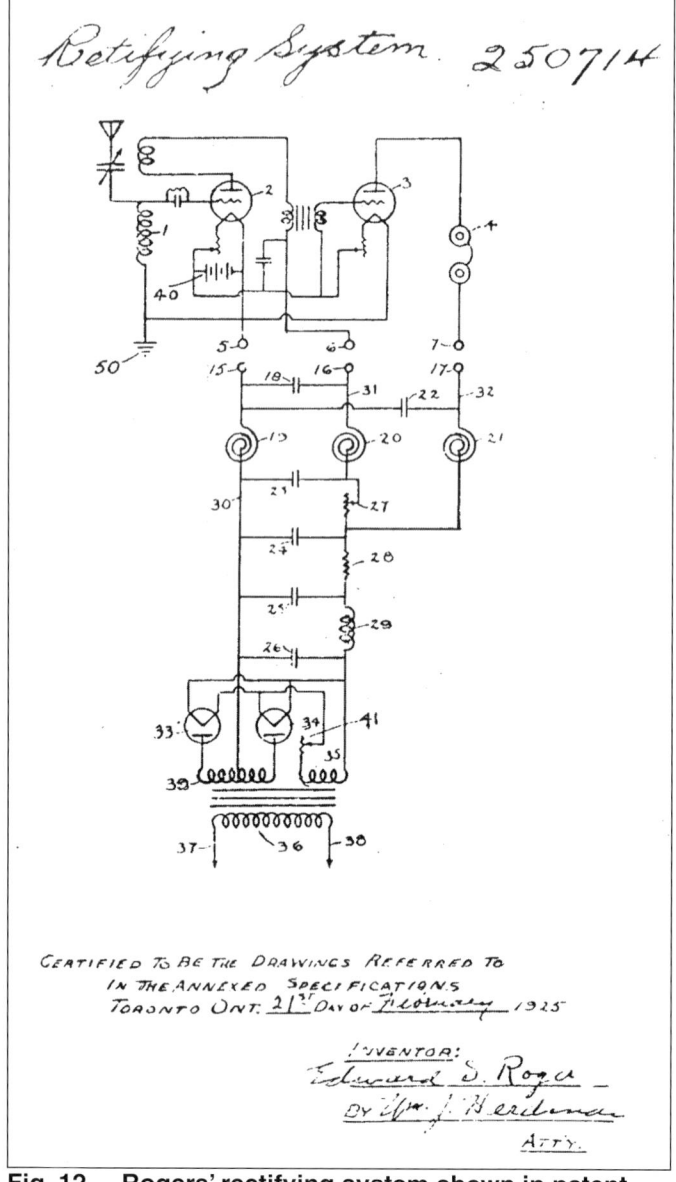

Fig. 12. Rogers' rectifying system shown in patent 250,714

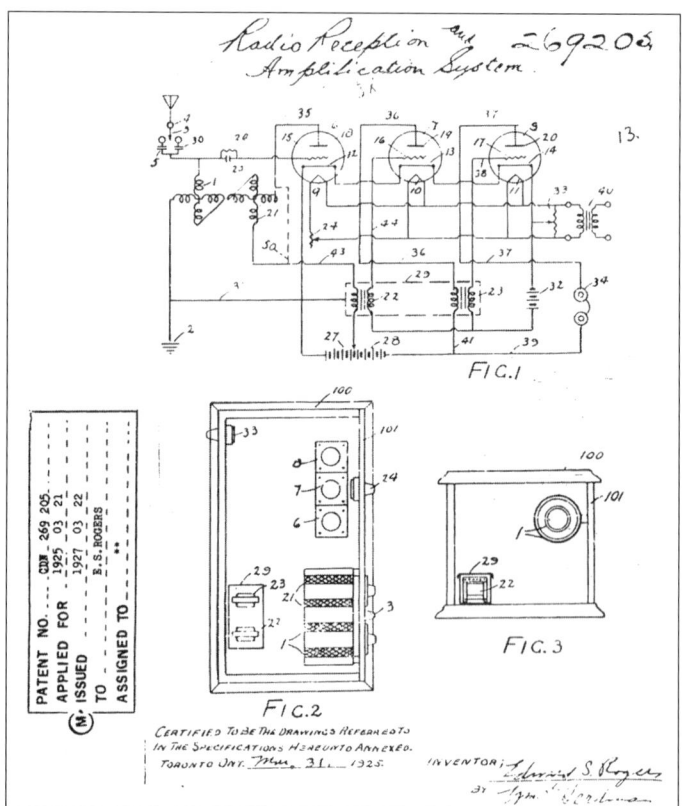

Fig. 13. Radio receiver with indirectly heated filaments shown in patent 269,205

Patented 1923-25. Other Patents Pending" (see Figures 16 and 17). Then, as more improvements to the tube were solely the results of Rogers pioneer work, the name McCullough was dropped.

During these months a number of prototype battery-operated receivers were made. A few of these were sold but most were equipped with the experimental AC tubes and sent out for field tests. A "B"-battery eliminator housed in a wooden case was also made and production samples of this unit were also sent out with some of the receivers.

The receivers were to be marketed under their designer's name, "Rogers". Coupled to this would be the registered tradename, "Batteryless". It was decided to operate the vacuum-tube and radio receiver operations as corporate entities so, on May 13, 1925, a new company, Standard Radio Manufacturing Corporation, Ltd., with an authorized capital of $500,000, was registered with the same directors and officers as Rogers Radio Ltd. The latter company concentrated on vacuum tube production. Together, the two companies had about twenty-five employees, most of them young, including a number who would later retire with over forty years service. Amongst these were J. R. (Bob) Eakins as vacuum-tube works manager, Stan Scott in the winding department, Jack Knapman in radio test, and Gordon Pipe as test engineer. The latter two were recruited from the Charles Branston company.

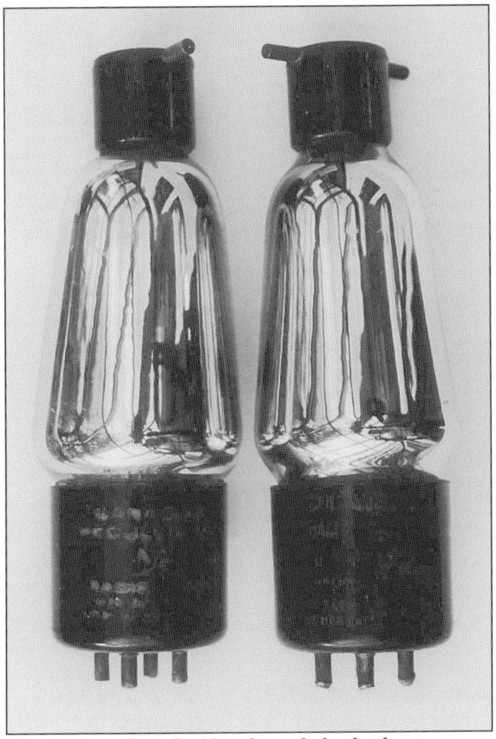

Fig. 14. Rogers prototype AC tube

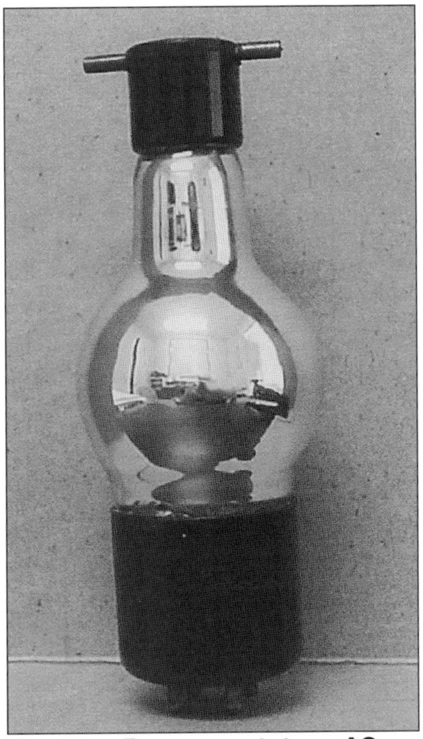

Fig. 15. Rogers prototype AC tube

Fig. 16. Original tubes labeled "Canadian McCullough AC Radio Tube"

Fig. 17. Cartons from the "Canadian McCullough AC Radio Tube"

QRS of Canada

The QRS Music Company of Canada Ltd., a subsidiary of QRS of Chicago, was registered in 1924 and opened showrooms at 690 King St. W., Toronto, early in 1925. Active in the player-piano and phonograph business, they had flirted briefly with the radio industry and were now acting as sales agents. As manager for the Canadian operation they were able to entice Fred Trestrail away from the Musical Merchandise Company. Fred was an aggressive, accomplished salesman with a flair for promotion. It was a good thing that he had radio experience since Musical Merchandise, in addition to distributing pianos and phonographs, were handling de Forest and Radiola receivers. Later, Fred with his brother, Burdock, would play an increasingly more prominent role in Rogers company affairs. This business relationship started, however, with the appointment of QRS as the Rogers distributors for Ontario and Quebec.[42]

The First Commercial Socket-Powered Receivers

Rogers tubes of an acceptable quality could now be made in quantity and full commercial production of the tube designated as type AC-32 started in August, 1925 (Figure 18). Some advance publicity was obtained when the radio columnist for the Toronto Star reported on August 18 being, a guest in a home "... where, with the aid of a new 5-tube tuned-radio-frequency receiver operating off the lamp socket and using no aerial or batteries, was enabled to tune in well over twenty-five stations".[43] Standard Radio, now with another Rogers, David, as works manager, readied receivers for display at the Canadian National Exhibition to be held in Toronto from August 29 to September 12, 1925.

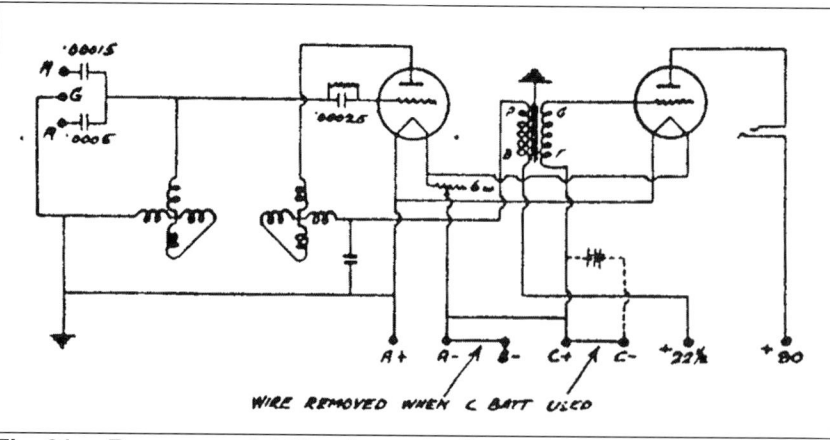

Fig. 21. Rogers model 20 schematic diagram

Fig. 20. Rogers model 20. A 2-tube battery-operated set with a regenerative detector and audio amplifier. Tubes and batteries were not included in the price of $38.50.

Fig. 22. Rogers model 30. A 3-tube battery-operated set with a regenerative detector and two audio amplifier stages. Tubes and batteries were not included in the price of $52.50.

Fig. 18. Rogers AC-32 (left) and Kellogg 401 (right)

THE ROGERS "BATTERYLESS" RADIO
is not only a Master Development — it is a PROVEN SUCCESS!

No "A" or "B" Batteries

No Aerial

"Just plug in - then tune in"

So great is the public interest in this newest advance in radio (the invention that utilizes electric current instead of storage batteries) that we have prepared a pamphlet explaining in the most simple manner the "WHY" of the Rogers Batteryless Set—how it operates and why its construction makes it so economical. Every intending owner of a radio set should get this pamphlet. Among other valuable information, this pamphlet tells why—

— the A/C Tubes in the Rogers Set have longer life than ordinary tubes in sets operated from batteries.

— why line voltage fluctuations cannot affect the operation of the Rogers.

— why the power required from your electric light system is less than the addition of a 60-watt lamp and the cost of current is less than 5c a week.

What About the Rogers Performance?

As the pamphlet explains the "WHY" of the Rogers Set, these remarkable testimonials tell about its success in operation:

(a) A letter from the Maritime says: "While our local broadcasting station was operating, we picked up 25 other stations on loud speaker without any interference. One was CKCL, Toronto. The first time this year that this station has been heard here."

(b) A Government Engineer, after severe tests, states: "Without hesitation, I would recommend this apparatus to the most exacting radio enthusiast."

(c) "In August last, under worst atmospheric conditions, secured transcontinental stations seldom heard in summer."

(d) A dealer writes: "The elimination of battery and tube troubles is a blessing both to the public and dealer. We have had wonderful results in tone quality, volume and selectivity. I can furnish you with names of many customers who are great Rogers boosters."

This New Pamphlet Answers All Your Questions -- Write for FREE Copy.

Whether you now own a radio set or not, you will want to read this specially prepared pamphlet, which covers in complete and concise terms every feature of this latest development in the science of radio reception. Write for your copy to-day. Address nearest distributor.

Rogers Radio Receiving Sets are manufactured under the DeForest Canadian Radio Patents, the E. S. Rogers Radio Patents, and the Canadian McCullough A/C Radio Patents by

STANDARD RADIO MANUFACTURING CORPORATION LIMITED, TORONTO
Owners of the De Forest Canadian Radio Patents.

DISTRIBUTORS -
Q..RS. Music Co. Canada, Ltd., 590 King St. West, Toronto - - for Ontario and Quebec
Radio Corporation of Winnipeg Limited, 290 McDermott Ave., Winnipeg - for Manitoba
Canada West Electric, Limited, Regina - - - - - - for Saskatchewan and Alberta
Radio Corporation of Vancouver Limited, 605 Dunsmuir St., Vancouver - for British Columbia

Fig. 19. Rogers Radio ad from *Radio News of Canada*, November 1925

Advertisements were placed in both the *Toronto Star* and *Globe* newspapers. A typical one reads - "Amazing Radio Set on Display at the EX. No batteries—No Aerial—Plugs into Any Electric Light Socket. 'Just plug in—then tune in' is the slogan. - A startling innovation in Radio is daily attracting crowds of Radio Fans as well as those to whom radio is more a mystery than ever. The idea of operating a radio set without batteries or aerial is almost uncanny. The new set certainly revolutionizes past ideas on radio operation, and intending purchasers of radio sets should see this one on display at the Standard Radio Mfg. Corporation's exhibit—and hear it in actual operation at the QRS Music Co. display-".[44]

Public interest generated by the Canadian National Exhibition exhibits was maintained with an aggressive advertising program in newspapers and magazines. Fred Trestrail of QRS was quickly able to build up a strong dealership network by offering closed territories and a rigid price maintenance structure. Three battery-operated models were among the

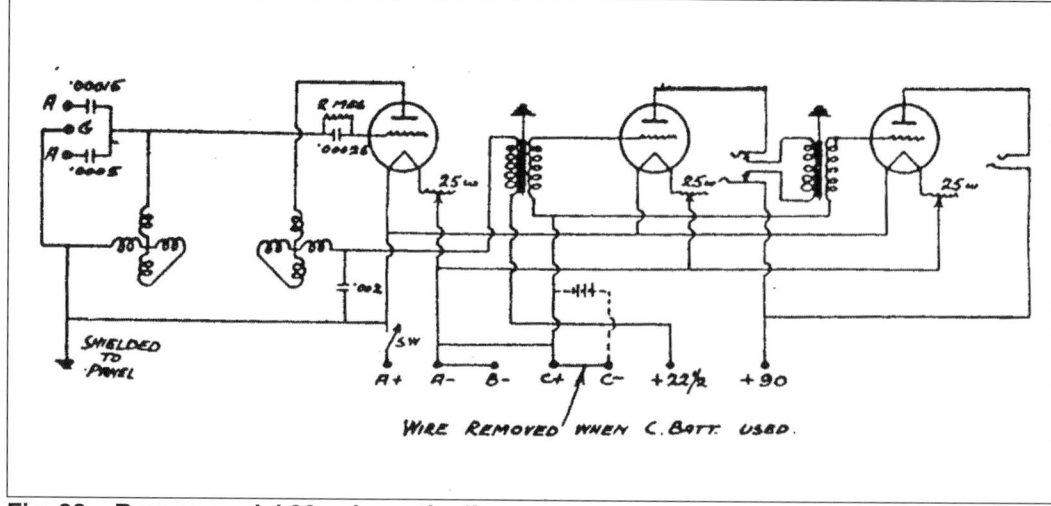
Fig. 23. Rogers model 30 schematic diagram

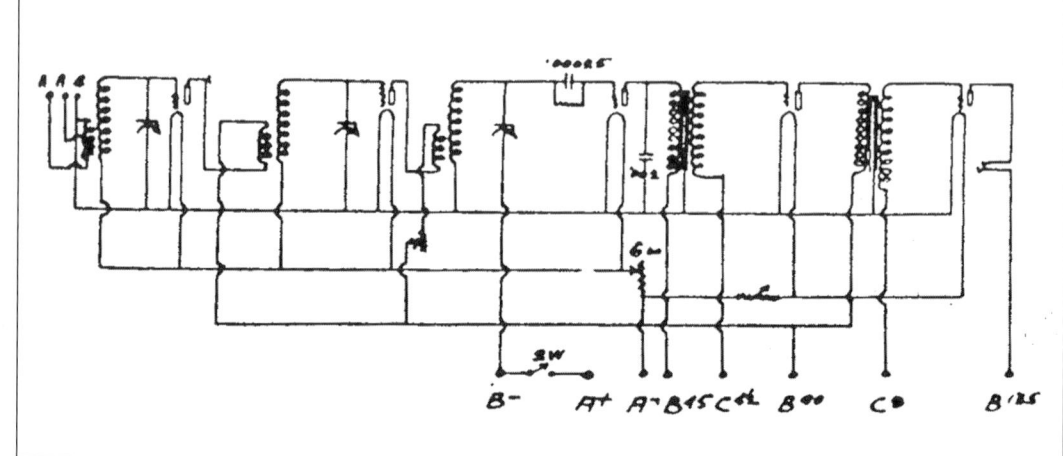
Fig. 25. Rogers model 50 schematic diagram

Fig. 24. Rogers model 50. A 5-tube battery-operated TRF receiver. Tubes and batteries were not included in the price of $130.00.

Fig. 26. Rogers B-battery eliminator using two Rogers R-100 tubes.

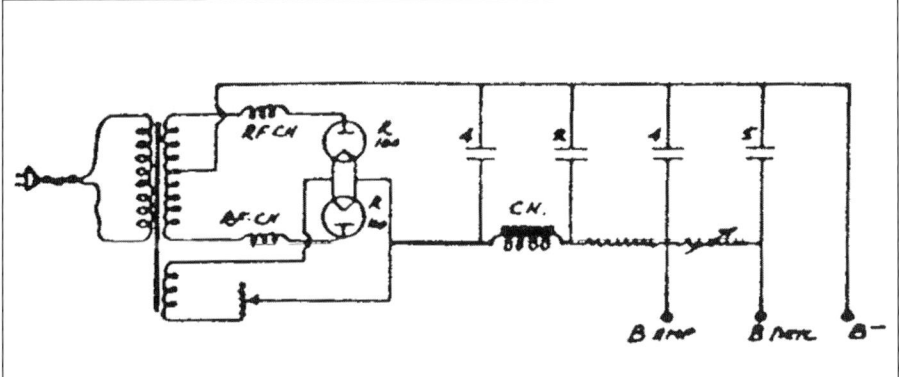

Fig. 27. Rogers B-battery eliminator schematic diagram

Fig. 28. Rogers model 130. A 3-tube set using Rogers AC-32 tubes in a regenerative detector and two audio stages. A Rogers type B heater transformer was supplied and the "B" supply was by batteries or "B"-eliminator. The "C" battery was behind the tube socket panel. Price with tubes and transformer but less batteries was $130.00.

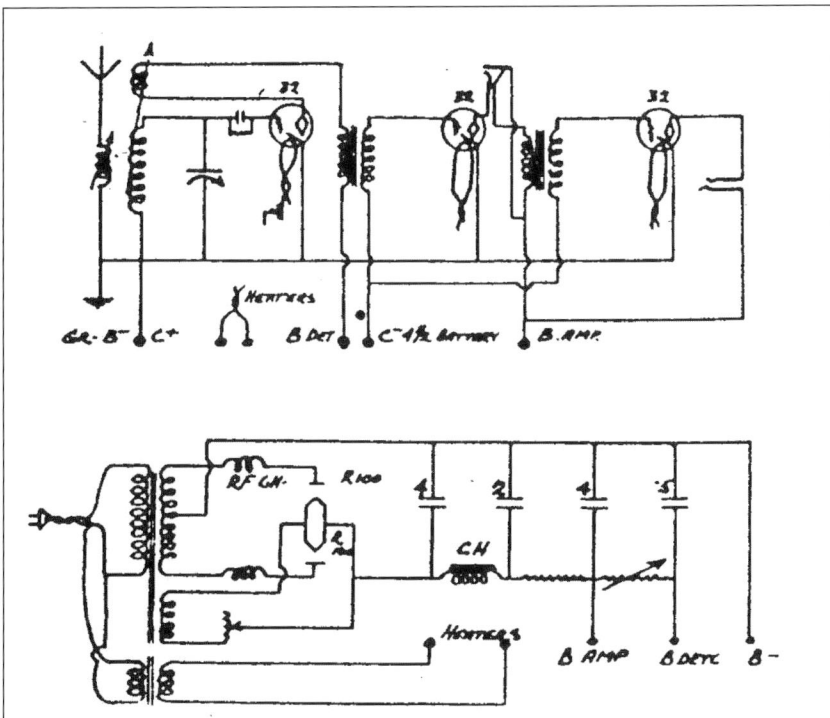

Fig. 29. Rogers models 130 and 135 schematic diagram showing "A" and "B" eliminators

eleven receiver types available. Prices ranged from $38.50 to $370.

Rogers 100-110-120 Circuit

The receiver depicted in the ad in Figure 19 was the model 110. The model 100 was the same except that it came with legs and amounted to a floor console. The model 120 excluded the internal loudspeaker (Figure 32). This 3-dial 5-tube TRF receiver resembles a Freshman with its honeycomb coils wound with green cotton covered wire and fastened with string to the back of each variable capacitor. Tube sockets for the Rogers AC-32 tubes were an integral part of the sub-panel. Volume control was provided by a damping rheostat across the 2nd. AF transformer secondary. A 400 ohm "stopper" resistor improved the first stage's selectivity.

The "A" and "B" power supply was enclosed in a sheet-metal box within the cabinet. The heater transformer had a multi-tapped secondary with the taps connected to positions 1, 3, 5, 7 and 9 on a selector switch (positions 2, 4, 6 and 8 were deadpoints to avoid shorting the secondary turns as the switch arm was rotated). Not shown on the schematic diagram was the 0-5 AC voltmeter monitoring the heater voltage. The "normal" 2.8 V operating point was marked and the meter was red-lined at 3.8 V. (Adjustment was necessary because of fluctuating line voltage). A "B"-supply voltage control operated by varying the rectifier tube's heater potential. It was first set about half-way then reset to bring the receiver to just below the oscillation point.

The instruction sheet on the inside of the cabinet lid stated that the set was manufactured under the Canadian de Forest, Edward S. Rogers and F.S. McCullough patents by Standard Radio Manufacturing Corp. Ltd.

1926-27—The Second Season

Additional manufacturing space was acquired at 858 Dupont St, Toronto,[47] and a second shift added to cope with customer demand for the receivers—a Niagara

Falls dealer reporting that his salesman was earning $50 per day in commissions on sales generated by door-to-door canvassing.

For 1926-27 the 3-tube receiver was unchanged except for its model number becoming 235 at $140 and the Model 130 was now the 230 at $150.

The R-100 rectifier tube was given a different filament construction and became the R-200. Two of these tubes were included in the new power packs used in the updated Model 120 and 110 receivers. Two "C"- potentials were available so that the below-panel "C" battery was finally eliminated to produce a true "battery-less" receiver. Additional "B+" decoupling was fitted to isolate the RF and output stages, and the 2nd. RF tube's plate supply could now be adjusted from a panel-mounted "sensitivity" control (Figure 34). The Model 120 now listed at $220.

Troubles were experienced with the R-100 rectifier tubes and all Rogers tube production was changed over to making replacements issued on a "no-charge" basis. The problems were traced to defective brought-in thoriated-tungsten filament wire. A change in supplier cleared the problem but it resurfaced three months later. Industrial sabotage was strongly suspected and the Rogers plants became very security conscious. Mullard's in England became the new supplier and no further trouble was experienced. Eventually all R-100 and R-200 tubes had the Raytheon gas rectifier as a recommended replacement.

Models 200 and 220 were introduced in the fall of 1926 (Figures 35 and 36). The 220 was a 5-tube plus rectifiers table model receiver in a solid-wood cabinet. Lifting the lid exposed three units: the RF section, the audio, and the power supply. Two AC-32 tubes acted as RF amplifiers with the three tuned circuits having their variable-capacitor pulleys ganged together with a metal belt to give single-dial control.

The tubular coils, one behind each capacitor, were mounted at the approved angle to minimize inductive interaction. A slow-motion vernier drive permitted more precise tuning.

The detector and audio stages were contained within a sheet-metal enclosure with the tops of the tubes protruding to allow for the overhead heater connections. A new type of power output tube, the AC-20 (320 mW) was fitted (Figure 37).

Fig. 31. Rogers model 135 was the same as model 130 but the smaller cabinet meant that both "A" and "B" supplies were external. Price with transformer and tubes was $110.00.

Fig. 32. Rogers model 120 was a 5-tube TRF in a solid walnut cabinet that was identical to the cabinet used on the model 50. Complete with tubes and power supply it cost $260.00.

Fig. 30. Rogers heater transformer Type B

transformer primary. The control knob for the line-adjust was located behind the rectifier tubes below the perforated grille (contrary to normal practice a clockwise rotation reduced the voltage). Not shown in the schematic is the AC voltmeter monitoring the tube's heater voltage. The meter face had been changed and has a "danger area" coded in red. The Queen Anne Console Model included a Baldwin magnetic speaker unit.

Naturally the increased sales of Rogers receivers affected the growth predictions of competing firms. They generally stressed the "more natural" reproduction available from battery operation and the advantages to be obtained from buying the products of long-established companies. On September 13, 1926, Canadian Westinghouse announced that their perfected batteryless receiver, using entirely new principles and standard Radiotron tubes, would be available for delivery on October 1. This was the Westinghouse Model 56 Desk type receiver retailing at $340. It used UV-199 tubes together with a UX-210 output tube with the filaments in series across the "B"-supply with the usual resistive-divider networks to obtain the correct operating points for the tubes.

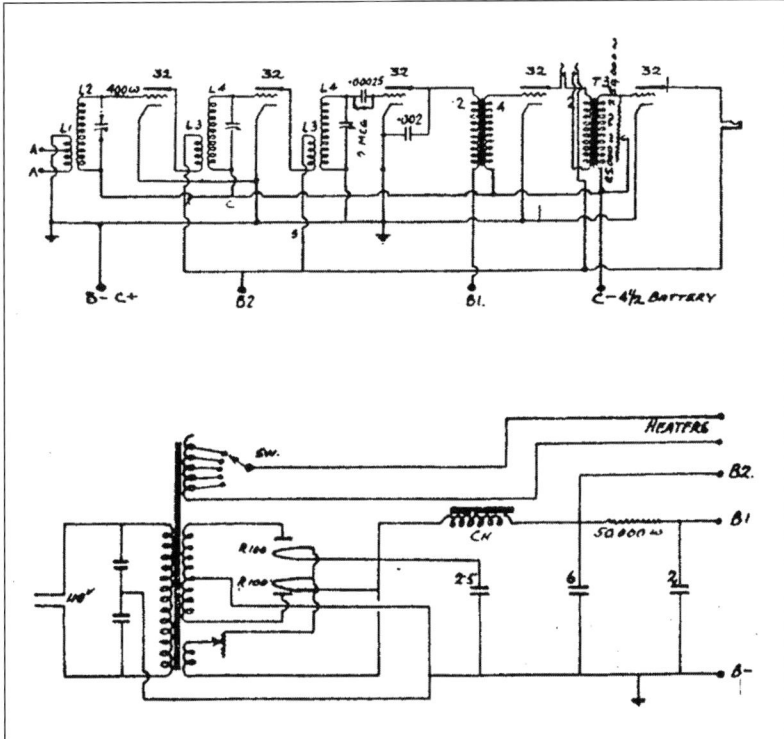

Fig. 33. Rogers models 100 - 110 - 120 schematic diagram

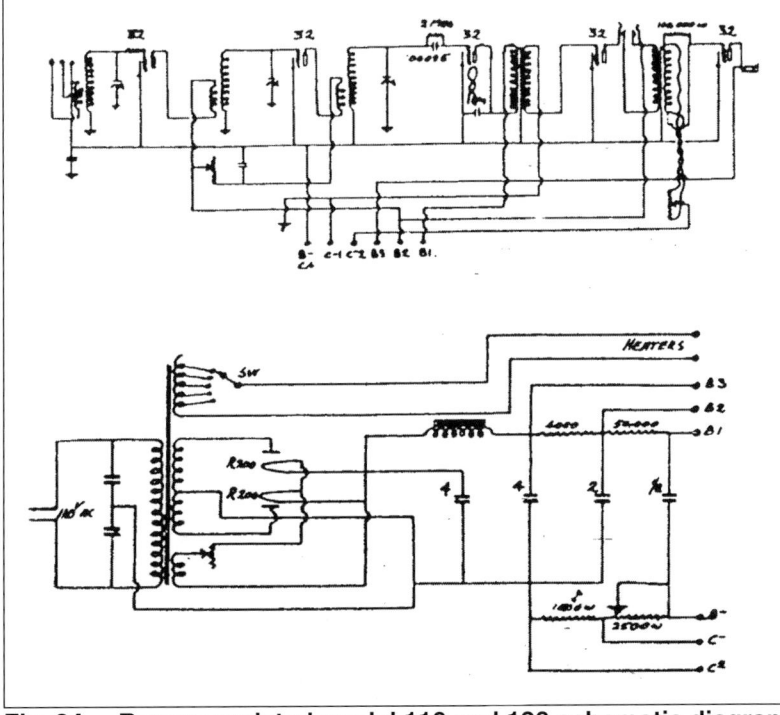

Fig. 34. Rogers updated model 110 and 120 schematic diagram in 1926-27.

Canadian Radio Patents Ltd.

1926 also saw an attempt to reduce possible costly patent litigation between the major Canadian radio manufacturers, Canadian General Electric, Canadian Westinghouse, Northern Electric, Canadian Marconi and the Rogers/Standard Radio interests. Each held patent rights which potentially could be in conflict. A new company, Canadian Radio Patents, Ltd., was formed in November and owned jointly by these major companies. Their patents were "pooled" into the company which then charged a royalty fee on each receiver sold by the manufacturer. Each year the revenue accrued in this manner was shared between the owners in proportion to the value of the patents contributed.

Radio Station CFRB

In addition to being involved with the design of the season's new receivers, Ted Rogers was realizing another ambition; to design, build and own a Toronto-based commercial radio station with a program content that would increase the demand for

The power supply used two R-200 rectifier tubes and additional filtering together with higher output voltages which were provided for the new output tube. The tap-switch heater control and the rectifier filament control were eliminated and a line-adjust control was connected in series with the power

radio receivers, particularly those made by Rogers. Suitable equipment was built at the Rogers plant, including the transmitting tubes which used hand-blown glass envelopes. The transmitter, located north of Toronto at Aurora (on part of the original Rogers land settlement granted in the year 1800), began testing in January, 1927 on l030 kHz. using 9RB as its call. Jack Sharpe was the engineer at the studio located in Ryan's Art Gallery on Jarvis Street, Toronto. On February 19, 1927, with Charles Shearer as manager, the 1,000 watt station, CFRB, was capturing its first radio audience. The RB in the call letters represented Rogers Batteryless.

Ted Rogers was elected to membership in the Institute of Radio Engineers in April, 1927 although he had been present with Gordon Pipe at the October, 1925 meeting at the General Electric plant when the establishment of a Canadian chapter of the IRE was first proposed.

The Third Season—1927-28

The Type U "B"-battery eliminator was introduced this season. Delivering 180 V. at 45 mA. It sold for $49.50 less tube. The Type 60 for 60 Hz. only, gave 135 V. at 35 mA. and retailed for $39.50 less tube.

Models 200 and 220 were retained from 1926 with some modifications to the power supply unit. A Raytheon BH rectifier was used and the filtering system was simplified (Figure 41). The model 220 was still $275.00 less speaker, with recommended speakers in the $25.00 to $40.00 range. A type 200 chassis was used in the new model 200-A. Originally listed at $395.00, the price of this drop-front console was increased to $420.00 in October, 1927.

The Model 90 introduced early in 1927 was a compact table model using a 5-tube TRF chassis with a two-dial tuning arrangement (Figures 42 and 43). An external power supply unit was required. A console version of this receiver with built-in loud speaker, "the Cameo", was available at $325.00. The chassis layout was very similar to the Model 120 with the detector and AF tubes in line at the back of the sub-panel chassis. The heater wiring was brought to the front of the panel. This in combination with the external power supply reduced the need for AF shielding.

The pleasing appearance of the Model 90 was not continued through to the Model 250, which was also introduced this season (Figure 45). The tube's heater voltmeter upset the visual balance of the front panel. This two-dial receiver had a self-contained power supply with the audio section again in a sheet metal shielding container on the right hand side. In a wal-

Fig. 35. Rogers model 200 was a 7-tube TRF receiver (5 tubes plus 2 rectifiers) that cost $395.00, tubes included.

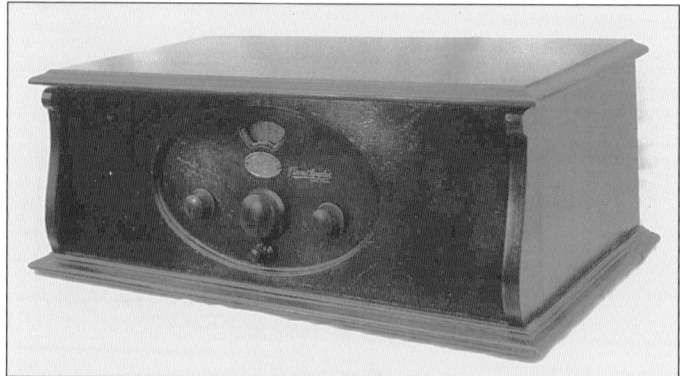

Fig. 36. Rogers model 220 receiver that cost $275.00 with tubes

Fig. 37. Rogers tubes AC-32 (left), AC-20, and AC-30 (right), all with etched tube numbers

Fig. 38. Rogers type 32 with paper label

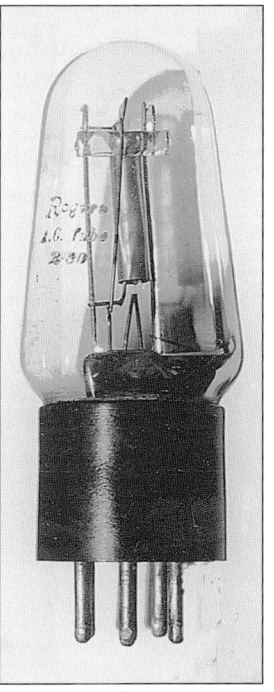

Fig. 39. Rogers type 30 with etched label

Fig. 40. Tube carton used in the 1930's crediting Ted Rogers with the AC tube on one panel and with his likeness on another.

nut cabinet, the retail price was $215.00 less speaker.

A. "Mel" Patience had joined Standard Radio as Chief Engineer at the end of 1926. A friend of Ted Rogers from university days, he had been working with Ontario Hydro and later would become Superintendent for RCA in Montreal.

The basic design of the Rogers receivers had not changed since 1925. In fact part of the appeal to dealers was the simplicity of the design. However, in 1927 the Jubilee chassis was introduced in the top-of-the-line sets which, in their Malcolm cabinets of walnut inlaid with maple, retailed for $890 (Figures 46 and 47). The receivers advertised as "...being developed by Ted Rogers over the last two years... the utmost in radio refinement and ability..." used a 7-tube chassis plus two UX-216-B rectifiers. The audio output stage used a directly-heated triode tube which originally was to be a type 171, then plans were made to develop a Rogers filament-type tube, the R-15 (Figures 48 and 49), but finally a UX-210 was used. Choke-capacity coupling was included to keep the DC plate current from the "Symphony" magnetic speaker unit. Four RF stages were put ahead of the detector with their variable-capacitors ganged for single-dial control. A bridge-type neutralization scheme was adopted with manual gain controls on the first and fourth RF stages. Only a limited number of these sets were sold and their performance was probably a source of disappointment to dealers and owners.

1928—A Year Of Change

In 1928 Rogers Radio Ltd. became the Rogers Radio Tube Co. Ltd., and a new single-ended indirectly-heated triode radio tube, the AC-30, was introduced

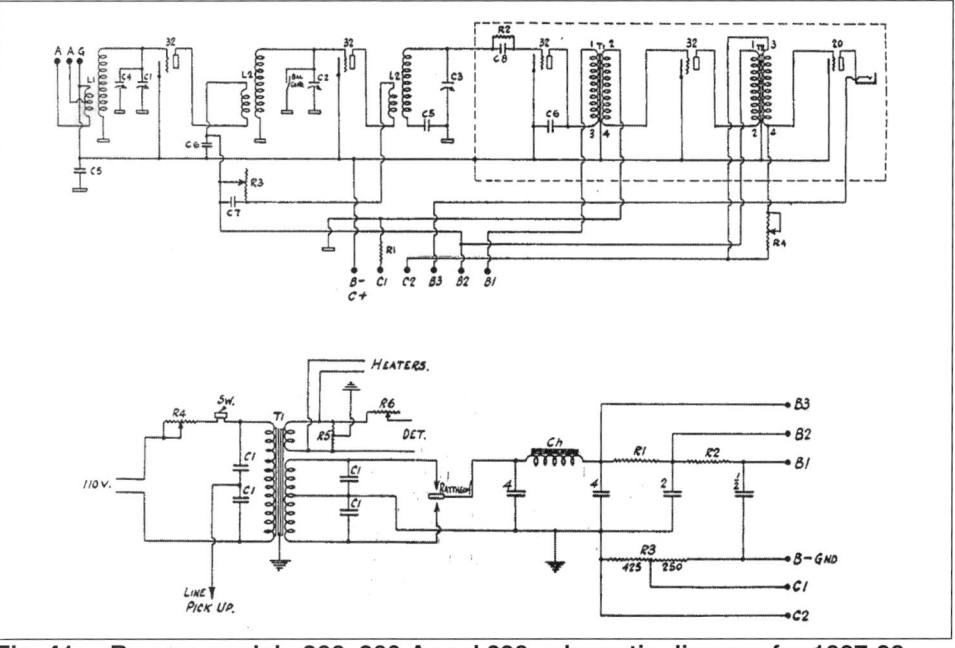

Fig. 41. Rogers models 200, 200-A and 220 schematic diagram for 1927-28.

(see Figures 37 and 39). The double-ended AC-32 tubes were included in the 1928-29 range of receivers. These were the Models 400, 420 and 400C which used the 'standard' chassis with the usual 5-tube TRF circuit, modified slightly by having self-bias on the RF tubes (Figures 50 and 51). The output stage called for the Rogers 15 tube but the 210 was used instead. The power supply used a Raytheon rectifier tube and 'automatic' line voltage regulation was obtained by using a Clarostat ballast unit.

The AC-32 tubes were also used in the 'advanced' chassis of Models 480 and 490 (Figures 52 and 53). This 6-tube plus type '80 rectifier receiver had three RF amplifier stages ahead of the detector and featured a complicated bridge-neutralization scheme for each stage with individual 2 k ohm variable resistors used for balancing. Both RF and AF gain controls were fit-

Fig. 42. Rogers model 90 receiver

Fig. 44. Rogers Commander model, a metal box version of the model 90 made for the Robert Simpson department stores.

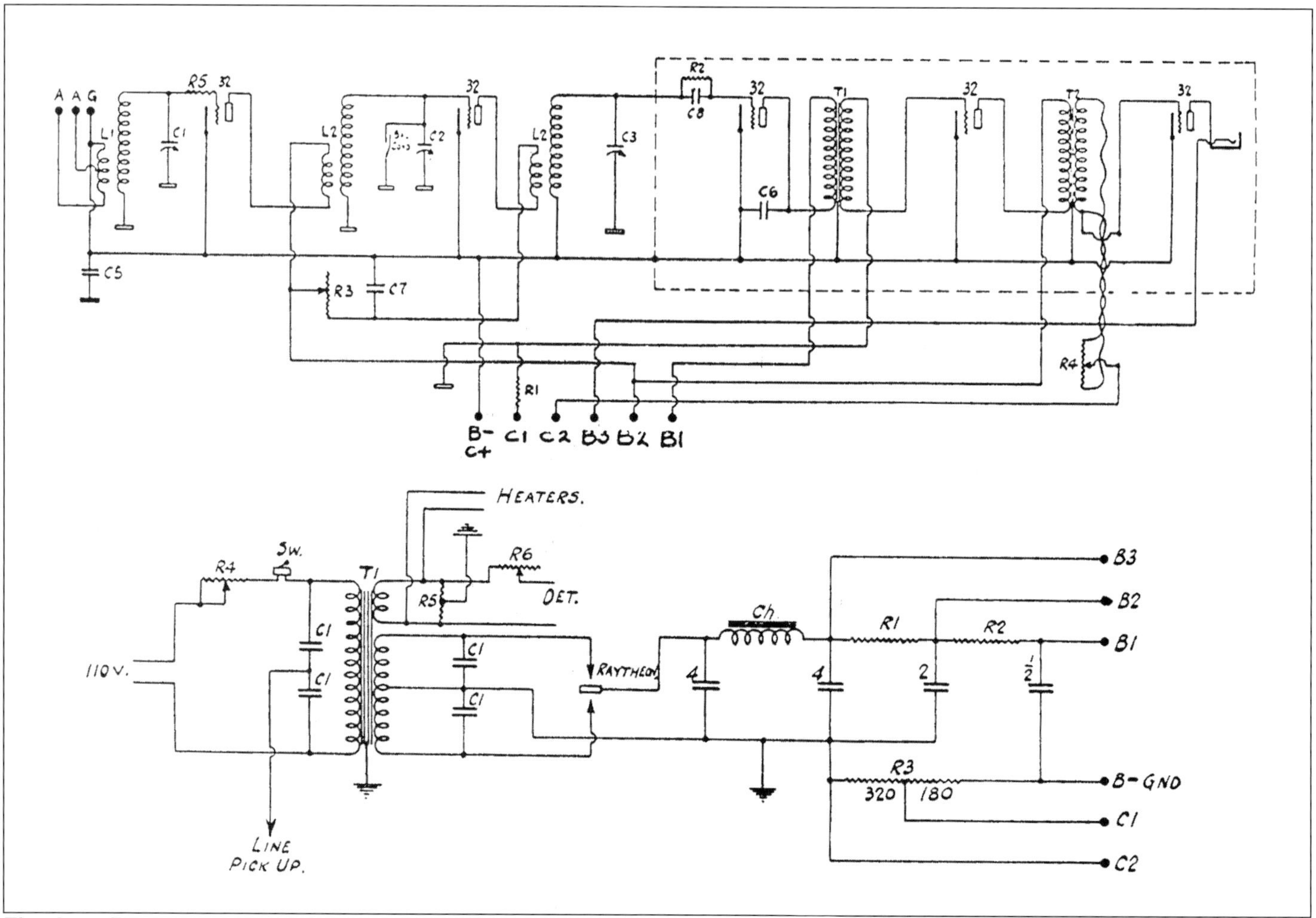

Fig. 43. Rogers models 90 and 250 schematic diagram

ted. Also included was a phonograph input jack.

A far greater change occurred in November when an arrangement was made with U.S.A. based Grigsby-Grunow, the largest independent radio manufacturer, to consolidate the Rogers and Majestic interests in Canada. The new company, Rogers-Majestic Corp. Ltd., replaced Standard Radio Manufacturing Corp. Ltd. Grigsby-Grunow acquired substantial stock and B. J. Grigsby was appointed to the Board. Albert S. Rogers, however, remained as Chairman. Majestic had recently purchased Phanstiehl for its RCA license and now obtained access to the patents, licenses and expertise needed as it moved into line-operated receiver production in the U.S.A. In Canada, Rogers was strengthened so that it could better withstand the competitive threat posed by the major companies now that they were incorporating the newly-available Westinghouse UY-227 tube in their receiver designs.

The involvement of Majestic brings to an end this first chapter in the Rogers Company history. The AC-20 and AC-32 tubes were obsolete. Henry "Hank" Parker arrived from General Electric and became responsible for tube development. The company survived the death of Albert Rogers in 1932, the collapse of the Majestic group in the U.S.A., and the depression. They took over the de Forest-Crosley interests in 1935 and were able to boast that they manufactured over 25% of the receivers sold in Canada.

Ted Rogers died on May 6, 1939. In 1941 all Rogers-Majestic assets, including patents, plant, equipment, etc. were sold to a British cable-cast company, Rediffusion, Ltd. Rogers-Majestic (1941) Ltd. became a subsidiary of Rediffusion (Canada) Ltd. All plants were now heavily involved in manufacturing military equipment. There was, of course, now no Rogers family involvement. Today, Rogers-Majestic is part of the North American Philips Organization. CFRB is the most successful independent Canadian radio station with Standard Radio Broadcast operating facilities located internationally and the Rogers Companies, with Ted Rogers Jr. at the helm, are active in electronic communication via cable and satellite.

Conclusions

It would appear that the development work for the AC operated radio tube should be credited to Westinghouse employees Freeman and Wade. Fred McCullough apparently evolved the manufacturing methods and Ted Rogers initially solved some problems that were preventing it from becoming a commercial reality.

A practical AC tube was, however, only a component, albeit an important one, in a successful line-operated radio receiver. To Ted Rogers should go the credit for recognizing the cause and solution for each problem that arose in the design and manufacture of the first commercial AC operated receiver. This, and his other accomplishments, make him Canada's foremost pioneer radio engineer.

Why did the Rogers company prosper while scores of others entering the field at the same time failed? First and foremost was probably Ted's

Fig. 45. Rogers model 250 receiver

Fig. 46. Rogers "Jubilee" chassis, 32.5 inches long by 16 inches deep

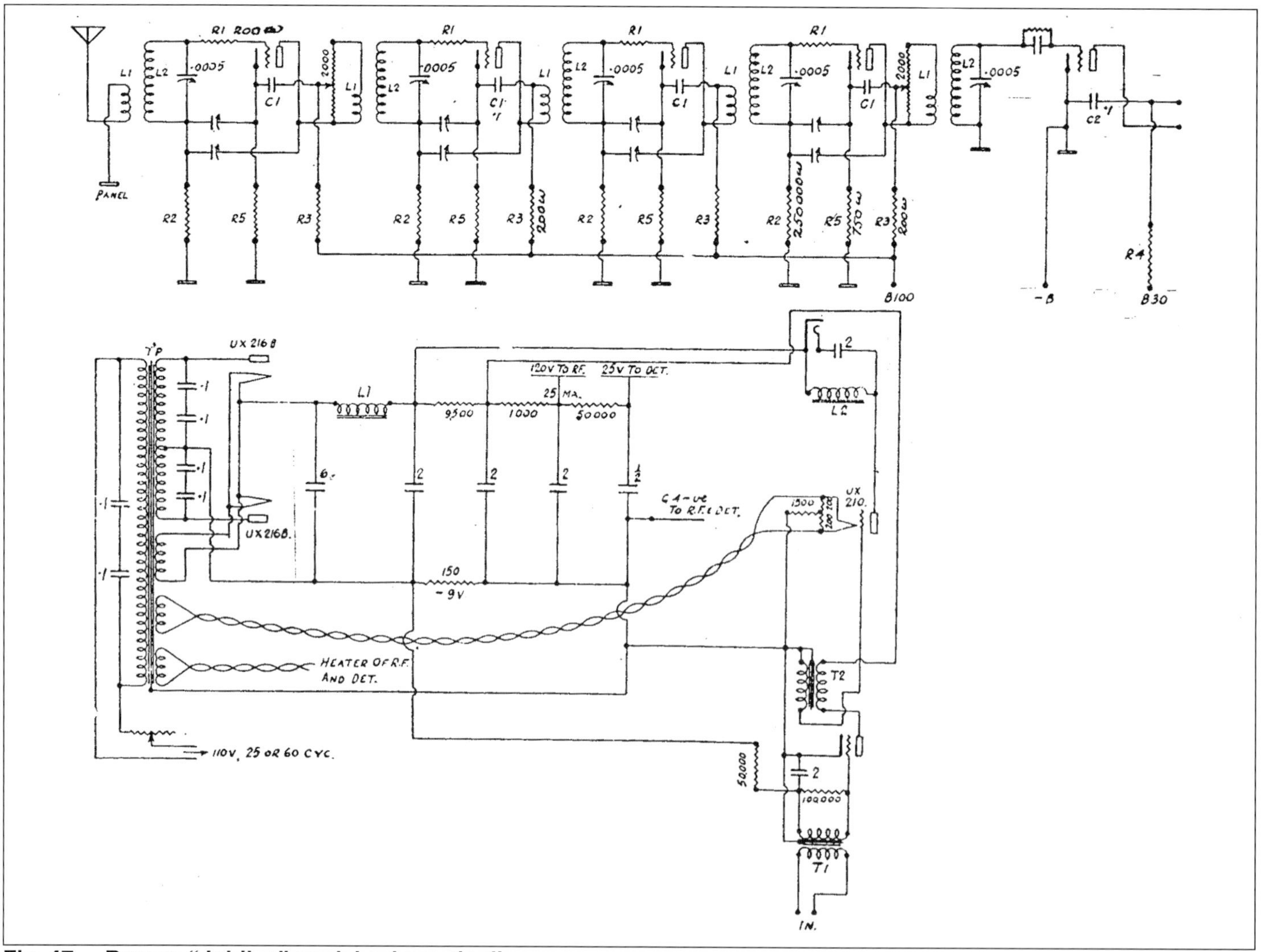

Fig. 47. Rogers "Jubilee" model schematic diagram

involvement with radio to the virtual exclusion of other interests. He had the self-reliance and persistence exhibited by many pioneer "ham" operators. His family background including not just their wealth and business acumen, but also their Quaker influence undoubtedly contributed to his ability, like Marconi, to inspire a band of devoted fellow engineers and employees. The product itself, its quality and performance, led to ready market acceptance. And finally, the aggressive marketing style of the QRS Organization led by the Trestrail brothers ensured a successful business outcome.

Fig. 48. Rogers tube type 15, which made a brief appearance in 1927

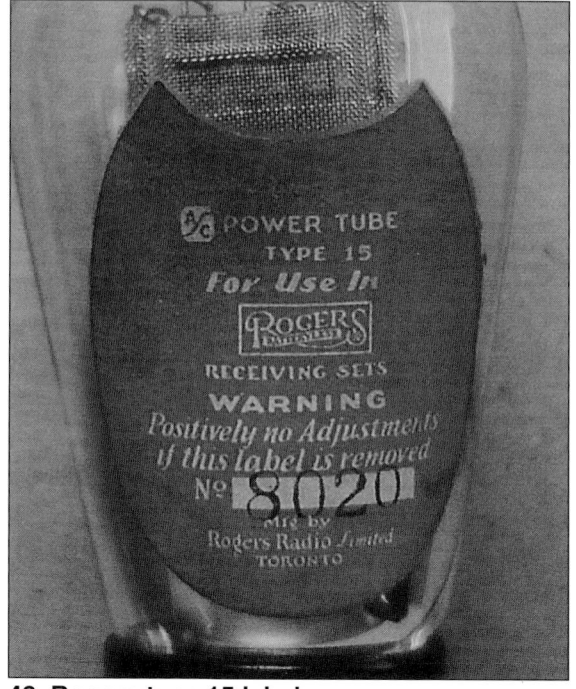

49. Rogers type 15 label

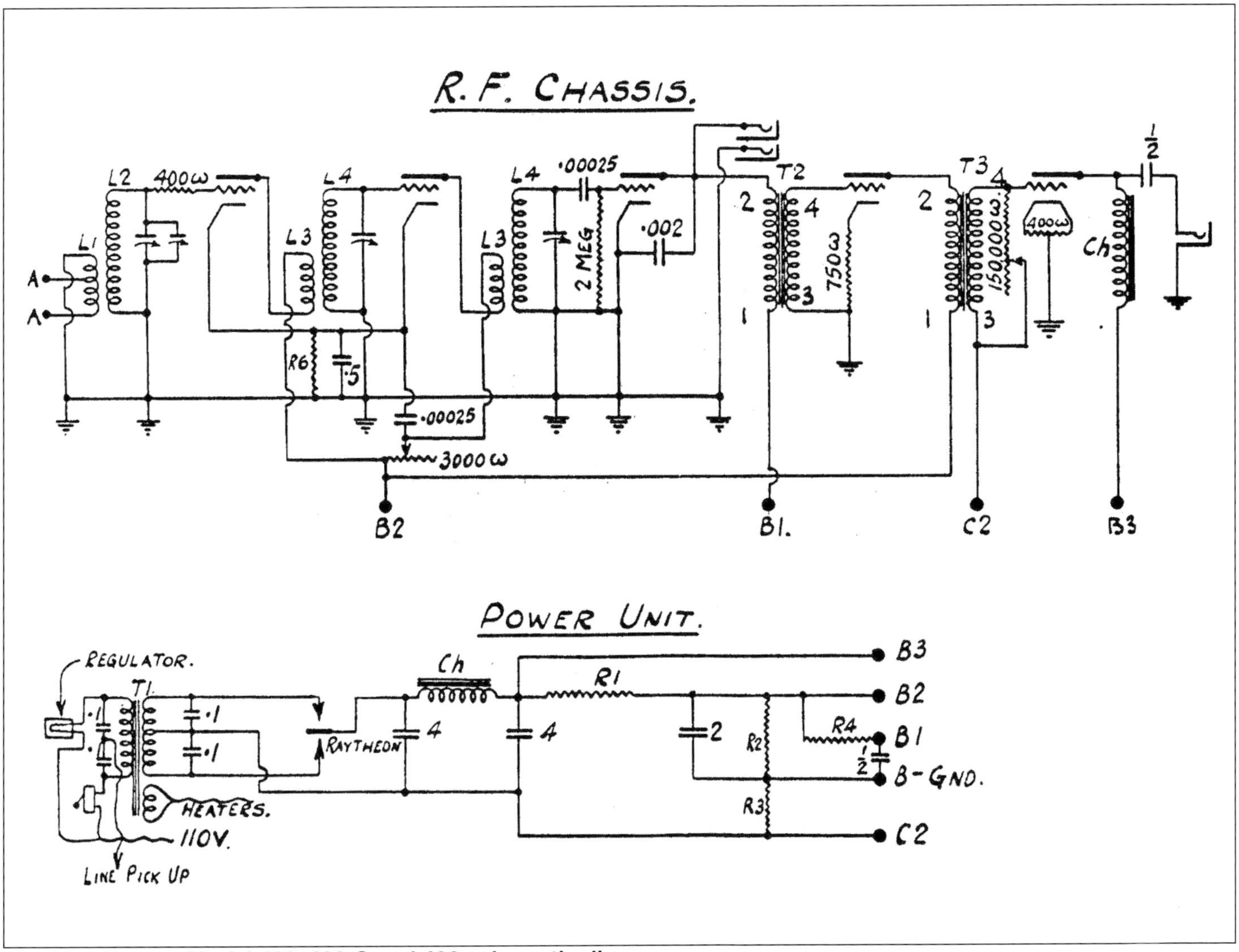

Fig. 51. Rogers Model 400, 400-C and 420 schematic diagram

Fig. 50. Rogers Model 420 receiver

Fig. 52. Rogers Model 480 receiver

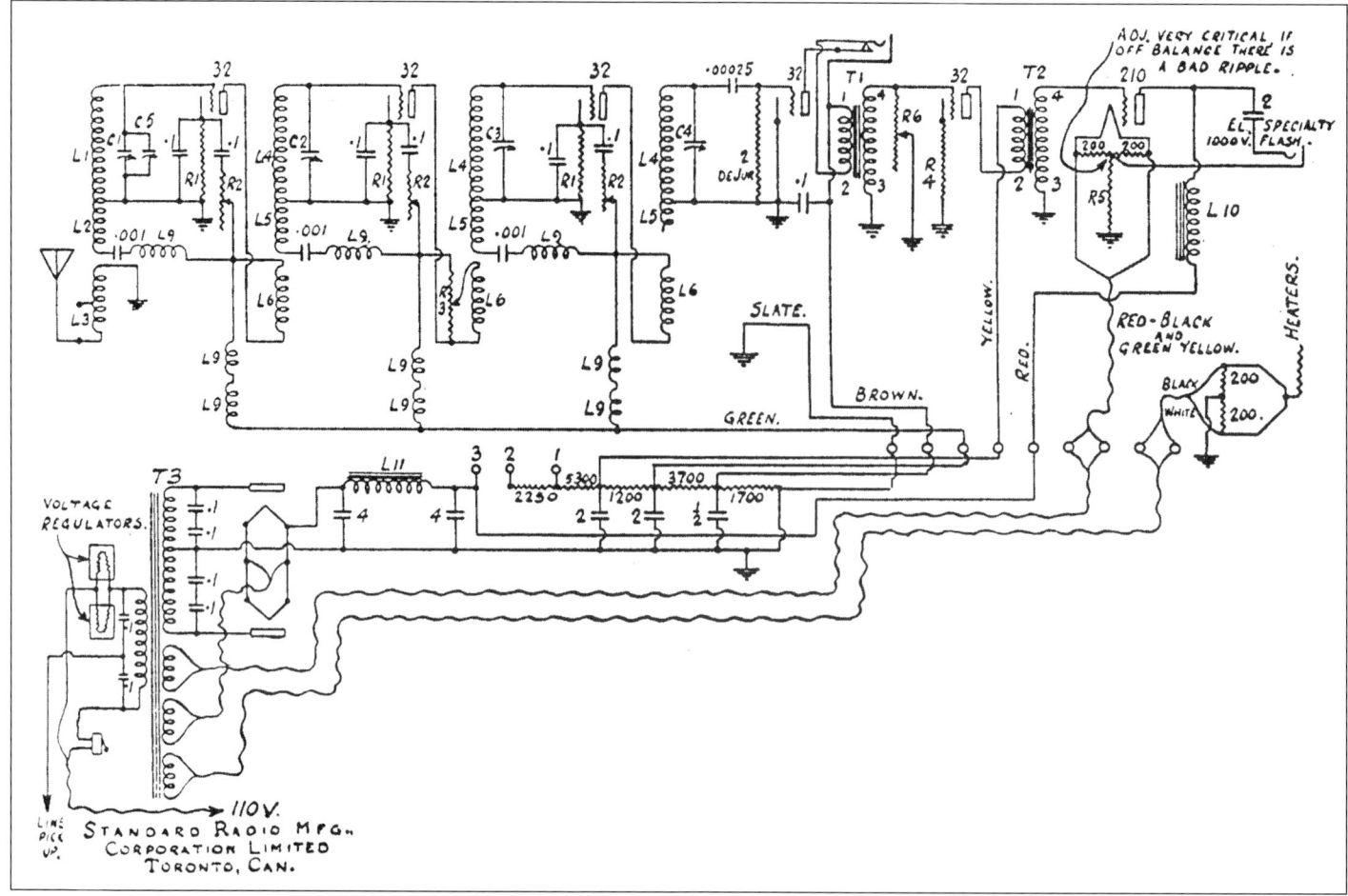

Fig. 53. Rogers Model 480 and 490 schematic diagram

Acknowledgements to the 1984 Edition

I wish to thank Aaron Solomon for his suggestion that this paper be prepared, M. Batsch for the loan of early radio tubes from his collection, Fred Hammond for access to his Museum, E. DeCoste and the National Museum of Science and Technology, J. Fawn for information on fellow Rogers employees, G. G. Armitage, Cdn. Region 7, IEEE, the Radio College of Canada, N. Burks of the Canadian National Exhibition, the Public Libraries of Toronto, Hamilton and Pittsburgh, Pa., and Joyce Chaplin for research assistance.

Acknowledgements to the 2002 Edition

The editors are indebted to Lloyd Swackhammer, first for his efforts in locating Maurice Chaplin after a 17 year absence from antique radio activity, and also for the provision of photographs of many of the Rogers artifacts in his collection. Robert Murray edited this manuscript for the AWA Review, and updated many of its illustrations.

Photo Credits

Line drawings are the originals supplied by Maurice Chaplin. Schematics and advertising are from Rogers company originals. Patents were searched by Maurice Chaplin. Figures were provided as follows: Doug Brighton, 15, 48, 49; Brian Darby, 42, 52; Stuart Johnston, 20, 31; Robert Murray, 2, 16, 17, 18, 22, 24, 28, 30, 36, 37, 38, 40, 45, 50; Lloyd Swackhammer, 10, 11, 14, 26, 32, 35, 39, 44, 46. Figures of artefacts now at the Canada Science and Technology Museum are: 24, 28, 30, 45.

REFERENCES

1. Lewis, T.S.W. Empire of the air: the men who made radio. New York, NY: HarperCollins, 1991, pp. 105-107.
2. See chapter 3.2 above.
3. See, for example, Willard battery advertisement, Radio News, August, 1922, p. 295.
4. Such a scheme was, however, used by Stromberg-Carlson in their 1927 model 523.
5. Gould 'Unipower'. See Adv. *Radio News*, September, 1923. p. 291.
6. The 'humdinger' was the subject of a 1919 French patent taken out by the 'Societe Independente de TSF'.
7. The French "S" tube of the Societe Ducretet was developed for this purpose in 1922.
8. The Garod model E-A, designed by B.F. Meissner, had the "B"-supply current passing through the detector tube's filament.
9. A.Wehnelt. German patent. DRP 157945. Jan. 13, 1905.
10. H. J. Round. British Pat. 6476
11. A. McLean Nicolson. Application April 16, 1915. Granted as U.S. 1,459,412, June 19, 1923.
12. H. W. Freeman. U.S. Pat. 1,291,641. Granted Jan. 14, 1919.
13. S. Torrisi. U.S. Pat. 1,368,584. Feb. 15, 1921.
14. Freeman and Wade. Cdn. Pat. 265,517. Issued Nov. 2, 1926.
15. Freeman and Wade. British Pat. 209,415. January, 1925.
16. Tyne, G.F.T. Saga of the vacuum tube. Indianapolis, IN: Sams. 1977. P. 319.
17. A. W. Hull. Paper presented to N.Y. Sect. I.R.E. Nov. 1, 1922.
18. A. N. Lucian. *Radio News*. May, 1926. P 1546. U.S. Pat. 1,677,977 (see also O.T. McIlvaine. similar U.S. Pat. 1,599,180)
19. F. S. McCullough. U.S. Pat. 1,403,700 Jan. 17, 1922.
20. F. S. McCullough. U.S. Pat. 1,496,243. June 3, 1924
21. Lived at five different addresses in Wilkinsburg and Edgewood Pa. during 1922-26.
22. F. S. McCullough. U. S. Pat. App. 128,375 Oct. 30, 1916.
23. Proc. I.R.E. Vol. 10. Pp. 468-485. December, 1922.
24. Cdn. Pat. 234,453. Sept. 25, 1923. (Appl. Sept. 1922).
25. Cdn. Pats. 244,433; 244,434; 244,435. Granted Nov. 11, 1924. Catkin valves were introduced in Britain, first as transmitting tubes with Cooled Anode Transmitting construction. The anode was a copper cylinder completely exposed to facilitate cooling. These appeared following World War I and some receiving tubes of similar construction were introduced in May, 1933. See J. W. Stokes, 70 Years of Radio Tubes and Valves, Vestal, NY: Vestal Press, 1982, Chapter 13.
26. E.S. Rogers. Cdn. Pat. 268,909. Issued March 8, 1927.
27. Premises were the National Carbon 'Ever Ready' plant during World War I, then the T. Eaton Co. warehouse. Now a parking lot for Toronto's downtown Holiday Inn.
28. Canadian Independent Telephone Co. headsets were still advertised by the T. Eaton Co. (for $2.98 a pair), September 3, 1925.
29. E. L. Koch. U.S. Pat. 1,677,896. July 24, 1928 (Appl. April 25, 1925)
30. F. S. McCullough. U. S. Pat. 1,677,900. July 24. 1928 (Appl. May 6, 1925)
31. F. S. McCullough. Cdn. Pat. 265,022. Oct. 12. 1926 (Appl. May 29, 1925)
32. Named after its President, G.W. King. Premises now demolished and the site of a Service Station.
33. L. Cockaday. 'How to build a 5-tube AC receiver'. Popular Radio, June, 1925, Pp. 511-523.
34. Morison Elect. Amplex. Haynes-Griffin. S. Hamner.
35. Dongan Electric.
36. Amplex, Freshman, PRS.
37. *Popular Radio*. January, 1926. p. 89.
38. *Popular Radio*. February, 1926. p. 59.
39. *Popular Radio*. September, 1926. p. 497.
40. Reported in *Radio Broadcasting*. July, 1926. p. 248.
41. Barawik Co. Summer Catalog. 1929.
42. QRS stayed as distributors until Majestic involvement in 1928. Fred Trestrail acquired control in 1930. Company became QRS Neon Corporation, Ltd. with David P. Rogers as V.P. Became Outdoor Neon Display, Ltd. with Rogers as Pres. in 1937.
43. The home at 37 Grenadier Road was that of Maurice Fiegehen, a draftsman, later listed as a Rogers employee.
44. *The Globe*. September 9, 1925.

RADIO AND PHONOGRAPH

A MODERN RADIO
Backed by a four year record of proven performance in batteryless operation and

A MODERN PHONOGRAPH
to play your favorite records with new beauty of tone by electrical reproduction

ALL IN ONE CABINET AT A MODERATE PRICE

ROGERS "FOUR HUNDRED" HIGHBOY PHONOGRAPH COMBINATION

Price Complete
$325
Slightly higher west of Fort William

AS A RADIO RECEIVER ONLY WITH PHONOGRAPH PLUG IN JACK

Price Complete
$250
Slightly higher west of Fort William

HERE is an instrument of exceptional entertainment ability. First of all a fine radio receiver of long established merit, providing proven Batteryless Radio Performance—and the added enjoyment of electrical reproduction of phonograph records. With the Rogers-Batteryless Radio Receiver, backed by a four year record of proven performance, has been combined a phonograph of modern design.

A simple turn of a switch transfers you from Radio to records.

ROGERS BATTERYLESS A/C
TRADE MARK REGISTERED IN 1925
RADIO AND PHONOGRAPH COMBINATION
The Complete Musical Instrument

THE Phonograph plays your records through the audio system of the radio. By this process a new beauty and tone quality is achieved. This electrical reproduction of records also enables you to control tone volume to suit your mood, from a whisper to concert hall volume.

See and hear this instrument and you will appreciate its full capabilities of musical expression— and the remarkable value offered by its moderate cost.

"Just Plug in— 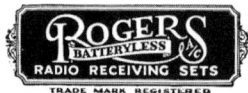 —Then Tune in"

ROGERS BATTERYLESS A/C RADIO RECEIVING SETS
TRADE MARK REGISTERED

Created and Manufactured Solely by
STANDARD RADIO MFG. CORP., LIMITED—TORONTO
Owning and Operating CFRB, Canada's First Batteryless Broadcasting Station

SOLD BY LEADING DEALERS IN EVERY COMMUNITY

~ 5 ~

Patents in Radio

Patent rights were complex in the field of radio, and at times there seemed to be as much legal activity in early radio as there was invention. To succumb for a moment to a vast oversimplification, the access to patents tended to be concentrated in the hands of larger corporations, not smaller ones.

The late Fred Hammond of Hammond Manufacturing Co. once told me that his father's business development was affected by this phenomenon. Hammond originally made broadcast receivers in about 1925. When patent lawyers took an interest in the fact that they did not have formal access to patents covering some of their work, the Hammond Company promptly turned to the manufacture of radio components. They were successful at this second line of business, and the Hammond Company is still in this business (and others) today.

This chapter defines patents and gives some examples of how they impact the radio industry. The next three chapters deal with smaller companies that confronted their inability to obtain access to crucial patents by several courses of action—retreat, compromise, or the provision of "kit" radios.

Patents and the Radio Industry in Canada: Developments to the End of the First World War
by Gordon Symonds

The legal manufacture of radio in Canada has relied almost totally on the availability of the patented inventions of many persons in the international community. The process of discovery, invention and patenting provide the historian with an excellent insight into the individuals who made the discoveries and the companies which exploited them. The story is a very interesting one, full of colourful individuals, human inventiveness, corporate and personal greed, xenophobic nationalism and political intrigue. When the legal profession is added to this mix, the result is indeed truth being stranger, and far more interesting, than fiction.

This article describes how radio patents have been dealt with in Canada, and provides essential background in the understanding of the patent 'situation'. It covers important developments up to the end of the first world war.

What is a Patent?
A patent is a grant from the Canadian government which gives an inventor the sole right to make, use and sell an invention for a period of twenty years. A patent holder then becomes the owner of an 'intellectual property', formed by the use of intellectual creativity. The right conferred by a Canadian patent extends throughout Canada, but not to foreign countries. Patents are granted for products, compositions, apparatuses and processes which are: New (first in the world), Useful (functional and operative) and Inventive (not obviously related to what was previously known in a specific area). A patent is not to be confused with: a Copyright (for literary, artistic, musical and dramatic works), Industrial Design (for shapes, patterns or ornamentation of an industrially produced object) or a Trade Mark (words, symbols, slogans or combinations of these which represent the origin of goods and services). One might describe a patent as being a 'reward for ingenuity'.

A patent owner may enter into agreements with others whereby in return for a royalty fee a license is issued to manufacture and/or use and/or sell the object of the patent. If a patent is infringed (used without the owner's permission) recourse is through the courts for damages.

In Canada, the governmental agency dealing with

patent matters is the Canadian Intellectual Property Office, a part of Industry Canada. The on-line patent database, at <patents1.ic.gc.ca>, contains information on all patents issued since 1920, information on earlier patents being available at the office in Ottawa for manual searches. Patent information is public information.

Since its beginnings in the 18th century and through the 19th century, with discoveries of the electron and electromagnetic wave radiation, the field of radio has been dominated by men of large inventive genius and insight, often with egos and temperaments to match. As in other complex scientific fields, no one individual can be said to have 'invented' the radio, rather it was a progression of development by many individuals in many countries over many years. One byproduct of this progressive development was an equally progressive international trail of patents filed by the inventors. Reginald Fessenden alone, for example, received over five hundred patents.

Notable Early Canadian Scientific Patents

In the latter part of the 18th century, the Canadian patent office recorded some noteworthy scientific achievements:

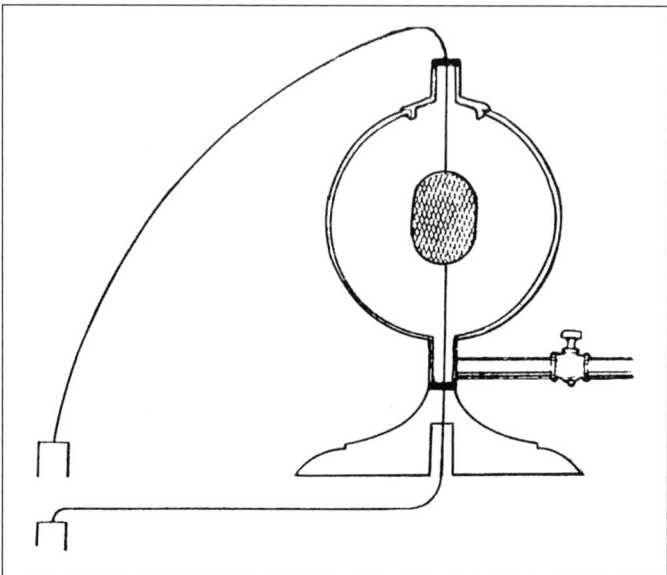

Fig. 1. Woodward & Evans' Electric Light, patent 3738

No. 3738. HENRY WOODWARD and MATHEW EVANS, Toronto, Ont., 3rd August, 1874, for 5 years: "Electric Light." (Lumière électrique.)
Claim.—The placing of carbons B, in lamps or other suitable vessels A, filled with rarified gas possessing the property of not chemically combining with the carbon when in a state of incandescence in connection with the arrangement of the electrodes E, E, fixing or connecting the carbon B, as set forth.

Fig. 2.

(1) An interesting and little known pair of Toronto inventors, Dr. Henry Woodward and Mathew Evans, received Canadian patent No. 3738 for an "electric light" in August, 1874[1] (see Figures 1 and 2). Dr. Woodward's invention was filed in the United States on January 4, 1875, predating both Edison and Swan! Indeed, it is very likely that Toronto saw electric light before Menlo Park, New Jersey, Edison's home base[2].

(2) After the Bell family moved to Canada from Scotland in 1870, their twenty-three year old son, Alexander Graham, worked with his father as a speech therapist for the deaf. In 1874-76, a scientific approach to their work, coupled with awareness of the electric telegraph and the development of a successful microphone, led to the invention of the telephone. Patents No. 7789, 26701 and 26710 describe the inventions which permitted his father to be the first to transmit a long-distance voice conversation over a telegraph wire in 1876. The call was from the family home in Brantford, Ont. to his fathers's insurance agent in Paris, Ont.

Closer to home, the electric stove, patent No. 35896, etc., by Ottawa electrical engineer Thomas Ahearn was used to prepare the first 'electrical' banquet at the Windsor Hotel in Ottawa in 1892.

Patents and the Legal Profession

Because patents have value, they have long been the subject of dispute and, over the years, patent lawyers have enriched themselves in the often protracted legal battles over ownership and infringement. It is worth noting that the receiver of a patent is not necessarily the inventor. Managers would often, for instance, patent the inventions developed in their laboratories by their staff.

Famous patent battles over the years have included: Hargreaves V. Arkwright (the spinning machine), Bessemer V. Kelly (steel making), Edison V. Swan (the electric light — but don't forget Dr. Woodward!), Bell V. Gray (the telephone), the Wright brothers V. Langley (the aeroplane) and de Forest V. Armstrong (the regenerative receiver). The outcomes of these disputes were not always evident at the outset. For example while it is widely believed that Armstrong (and probably others) discovered the regeneration principle, de Forest's patent claim eventually prevailed in the courts.

The Beginning of the Radio Era (1895 – 1918)

The period from about 1895 to the end of the first world war saw unimagined advances in the field of wireless communications. After the validation of Maxwell's 1873 mathematical theory of electromagnetic waves by Heinrich Hertz in 1895, there was a rapid succession of fundamental developments, exploitations, inventions and patents: tuning (Oliver Lodge), the antenna and magnetic detector (Guglielmo Marconi), amplitude modulation and the heterodyne principle (Reginald Fessenden), RF alternators (Ernst Alexanderson) and RF arc generators (Valdemar Poulsen).

A Landmark Radio Patent — the Fleming 'Oscillation Valve'

An observation made by Thomas Edison in 1883 during his work on improving the light bulb (the so-called 'Edison Effect') went unrecognized and unused for twenty-one years until, in 1904, John Ambrose Fleming, a technical advisor to Marconi's Wireless Telegraph Company, filed his famous patent for the 'oscillation valve' (Figure 3).

The drawings for the 1883 patent filed by Edison clearly show a two element vacuum tube (Fig. 4 in the patent drawing which is also Figure 4 below), but Edison's patent application was 'deficient' in the fact that it did not make appropriate patent claims. Even if you patent a device, if you do not make appropriate claims regarding its operation, the feature not claimed is not protected. Unfortunately for Edison, he was a devotee of direct current, indeed it was one of his companies (a forerunner of Consolidated Edison), which was responsible for the use of 110 volt DC on the U.S. east coast. Accordingly, AC was considered by him to be a nuisance and he failed to appreciate the value of his device, which would convert 'the despised alternating current into well-regarded direct current'[3].

This device which Fleming patented we now call the thermionic diode. Fleming's invention was initially patented in England, the United States and Germany, with the U.S. patent being assigned to the Marconi Wireless Telegraph Company of America.

This seemingly arcane invention by Fleming was to become the basis of a world-wide industry and the details of its discovery and patenting were to have very important consequences indeed for radio in Canada.

Fig. 3. Fleming's 'Oscillation Valve' (British Science Museum Photo)

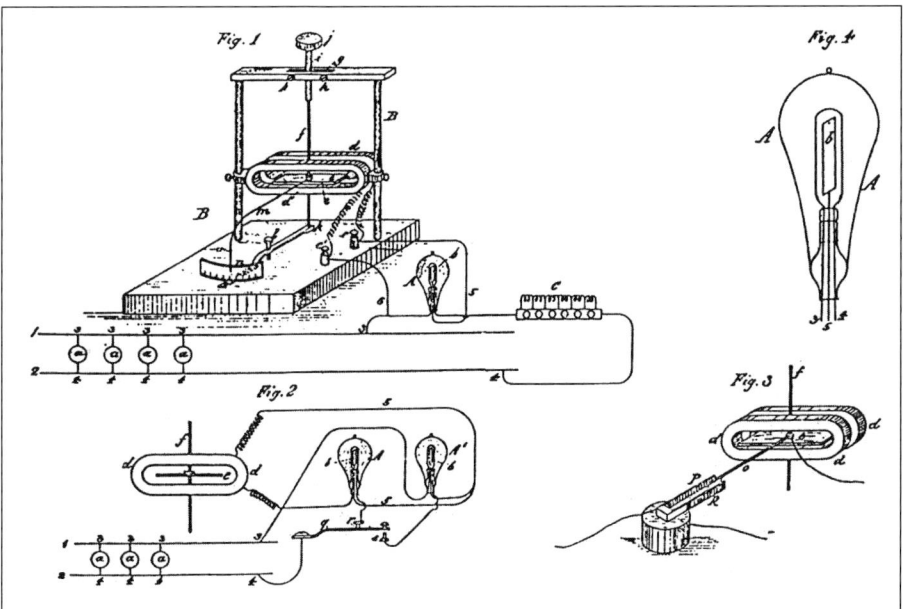

Fig. 4. Patent Drawings for the 'Edison Effect', 1883

Let the Litigation Begin!

One of the remedies available to a patent owner if his ideas, etc. are used without permission is to sue in the courts for infringement; one of the more litigious individuals in this regard was Lee de Forest of the

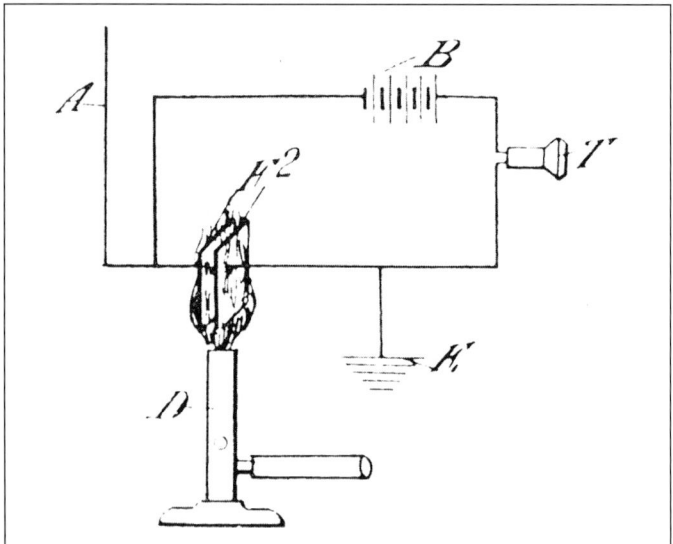

Fig. 5. De Forest's Flame Detector

United States. Prior to the development of Fleming's thermionic diode detector, he had developed and patented a diode-type detector using the ionization in a gas flame (Figure 5).

Then "by a series of patents (he) transformed his Bunsen-burner arrangement into a device which was virtually the same as Fleming's diode, and to which he gave the name 'audion'"[4]. This ruse, known as 'engineering around', did not succeed however and after a long legal battle Fleming emerged victorious in 1916.

In the interim, in 1906-07, Lee de forest made a real discovery equal in importance to Fleming's, namely the addition of a control grid to form an 'audion' (thermionic triode) which could amplify signals at radio frequencies.

Gridlock?

With the end of the de Forest V. Marconi litigation over the diode patent, the next link in the chain of events was a decision in the U.S. courts that de Forest's triode infringed Fleming's diode patent. Recalling that Fleming had assigned his patent to Marconi in the United States, this decision created what might literally be called 'gridlock'. The following two position statements, made at the request of the editor of an amateur radio magazine in 1917[5], outline the situation very well:

(1) Marconi Wireless Telegraph Company of America

Sir:
I have your note of inquiry dated 25th inst.
You understand that the Marconi Wireless Telegraph Company of America owns the United States Letters Patent issued to John Fleming, No. 803,684, November 7, 1905; that suit was brought against de Forest Radio Telephone and Telegraph Company and resulted in the decision of His Honour Judge Mayer in the Southern District of New York. In his answer in that suit, de Forest set up a dozen of his patents for improvements in detail, among others, two patents covering a third element in grid form located within the vacuum chamber and between the hot and cold element. That is the device bearing the name of 'audion'. The Marconi Company thought one or both of these patents were good and confessed judgement. Judge Mayer's decision sustained the Fleming patent and found that it was infringed by the so-called three element or audion type. Subsequently, an injunction was issued preventing de Forest from making, selling, or using any vacuum detector, and an injunction was issued upon de Forest's application, preventing the Marconi Company from making, selling or using valves having three elements within the vacuum, like the audion. The Marconi Company's injunction was modified to allow de Forest to make and sell the audion, provided de Forest gave suitable bond, rendered a periodic statement of goods made or sold, promptly appealed from Judge Mayer's decision, filed the record on appeal before the date set, and perfected the appeal without delay. De Forest made statements for a while, but failed to comply with the conditions here named and the Marconi Company's injunction immediately became effective.

The situation now is that de Forest cannot make or sell any valves for radio use and the Marconi Company cannot make the audion.

In this condition of affairs, the Marconi Company has decided to make a valve that does not infringe the de Forest audion patents and that is equally efficient and will do anything that the audion does, or is alleged to do. This valve will be placed on the market as soon as arrangements are completed, and will be accessible (sic) to amateurs. I am unable to tell you the exact date when the Marconi Company will be ready, but unless some obstacle now unforseen or unexpected presents itself, the date is not far in the future.

I send to you, under separate cover, a printed copy of Judge Mayer's decision, for your information.

Yours very truly
W.B. Vansize, Patent Attorney

(2) DeForest Radio Telephone and Telegraph Company

Replying to your letter of 25th inst. regarding the patent situation concerning sale and manufacture of Audions, would state that the injunction of October 9th, issued by the U.S. District Court was restrained later by the court on this company furnishing a bond, and also on the understanding that our attorneys would move for preference in the Court of Appeals.

For several months, therefore, we continued manufacturing and selling this apparatus, but on the 2nd of January our attorneys advised us that in order to gain time so they could more thoroughly prepare their case they would be unable to make the motion for preference, and that the injunction therefore would have to go into effect. Rather than jeopardize a favorable decision for us we preferred to sacrifice a few months sales on this class of apparatus, namely Audion and Ultra-audion apparatus.

For the present, therefore, we cannot manufacture or sell such goods. This does not effect the Amplifier bulbs, or the Oscillion transmitter bulbs.

The case will probably come up before the Court of Appeals towards the end of February, and we hope to get a decision within four or five months.

Very truly yours
Chas. Gilbert, Treas., de Forest Radio Tel. & Tel. Co.

The situation described above was obviously untenable. In the radio patent industry, the term used is 'interlocking patents', the most important feature of which is that one or both patents are worthless without the other.

Since the basic idea of patent legislation, such as the Canadian Patent Act, is to promote the public good, most or all national legislations in this field contain a very powerful and little known provision to discourage intransigence: regardless of who owns the patent(s), the invention(s) must be made available to the public. If a patent owner should assume a 'dog-in-the-manger' attitude and decide not to manufacture or license others to do so, then the legislations decree that his patent shall be cancelled! This is precisely what happened to the Radio Corporation of America in the United States — it refused to license others to produce the super-heterodyne circuit until, in the early 1930s, the courts gave them the option of doing so or losing their patent rights. The courts did not regard their

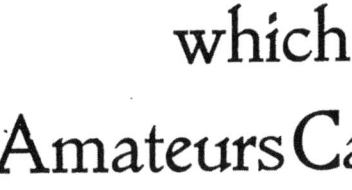

Fig. 6. Marconi Vacuum Tube advertisement from *QST*, Sept. 1919

token sets in the mid-1920s to be a good faith attempt to provide super-heterodyne receivers to the public.

The gridlock situation described above obviously could not continue, or eventually both Marconi and de Forest would lose their patent rights and vacuum tube manufacturing would become a free-for-all. Given the amount of money at stake (it would be similar to fibre optic patents in the current communications technology), no businessman could permit this to happen. Given the size of the egos involved however, it just might!

Apparently, Marconi made good on its word to produce some sort of vacuum triode 'that does not infringe the audion patents', for the following advertisement appeared in the September, 1919, issue of *QST* magazine (Figure 6). Within five months, in February, 1920, this advertisement had been revised to that shown in Figure 7.

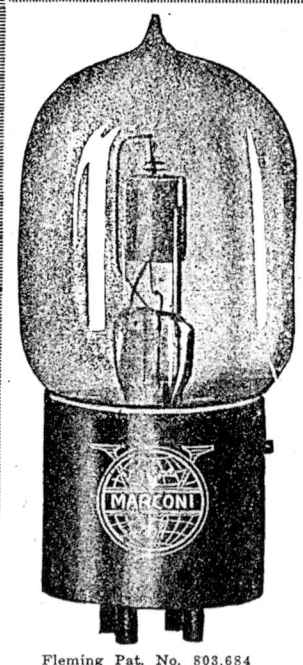

Fig. 7. Marconi Vacuum Tube advertisement from *QST*, Feb. 1920

REFERENCES

1. *The Canadian Patent Office Record*, August, 1874
2. *Canadian Electronics Engineering*, March, 1985, pp. 42-43.
3. Desmond Thackery, Edison's Electrical Indicator, *Wireless World*, February, 1984.
4. Vivian J. Phillips, *Early Radio Wave Detectors*, London: The Institution of Electrical Engineers, 1980.
5. *QST*, March, 1917, pp. 16-17.

~ 6 ~

W.W. Grant Radio

In 1991, when I expected to be in Calgary on business, I contacted Gordon Wilson in Edmonton and asked if there was anyone I could visit in order to learn more about W.W. Grant radios. He referred me to Ed Burge, a radio history enthusiast originally from Montreal who had planned to write a Grant company history. Ed had interviewed former employees, Grant family members, and collected newspaper clippings. He was elderly and frail when I met him, and doubted whether he would complete his book. He decided to lend me his documents from which I wrote the following article. He died a short time later.

"The Voice of the Prairie"
A Brief History of W. W. Grant (1892-1968)
Robert P. Murray

Litigation

William Walter Grant had been in the British Royal Air Force in World War I, and had been second in command of a radio and photographic installation and repair establishment near the front lines in France. He and his subordinates were installing and maintaining all the latest wartime radio equipment, and adjusting and improving it as they did so. When Grant returned home after the war, he went into the radio business.

Legend has it that Grant and H. Gordon Love were once friends, but that the friendship ended with Grant chasing Love down a Calgary street with a piece of 2 by 4. Love had become the local manager for Canadian Westinghouse, and Westinghouse had the Canadian rights to the Armstrong patents. Grant had sold one of his newly manufactured broadcast receivers to Love, who apparently had then advised his new employers at Westinghouse that their patent rights were being violated in Calgary.

There followed a dispute between Grant and Love that was charged with a great deal of passion in the mid-1920's. From this distance across time, it seems to have been essentially due to two events: First, Armstrong was slow in getting his regeneration patents registered in Canada, and second, Grant had come to believe that on the basis of his wartime work he had access to radio designs with which he

Fig. 1. W.W. Grant

had become familiar. In the early 1920's he was actively engaged in manufacturing devices which were covered by Armstrong's U.S. patents. He later freely admitted this in court. In the early days of his manufacturing though, he may not have been particularly aware of Armstrong's work.

After W.W. I, Canadian (i.e. British) law allowed a grace period of five years for patents to be registered on inventions where the war would have interfered with the usual registration process. On July 8, 1920 Edwin Armstrong applied for registration of his patents in Canada. They were granted on March 7, 1922, under patent number 216,321. By this time his U.S. patents were held by R.C.A., and in Canada the subsidiaries of the R.C.A. partners were assigned the patent rights.

Grant, in his war service, had operated a radio installation and repair depot for the R.A.F. The state of radio development being what it was at the time, technicians were obviously called upon to be creative in making whatever field improvements they could to the operation of the equipment. Grant was regarded as technically skilled by his peers, although he apparently had no formal training in engineering. He was responsible for a large number of service personnel doing installations and repairs (he said 300 men in an advertising brochure in 1922), and he was commissioned Second Lieutenant in the field in 1917. He claims to have made some improvements to the functioning of radio circuits for which he was responsible, and this would seem very plausible given the circumstances of his work.

The only invention for which documentation remains is referred to as the "clapper brake". For it he was awarded the sum of fifty pounds by the British government after the war. In his court testimony, Grant claims that the invention is inappropriately named, and that it is in fact a circuit improvement to enhance the selectivity of receiver tuning. After the original publication of this article in 1992, collector Louis Meulstee in Holland wrote to say that the clapper brake was evidently a device that improved the output efficiency of a WW I British aircraft spark transmitter which in turn enabled a number of these transmitters to operate simultaneously at different frequencies. Meulstee referred to an unnamed article in *The Wireless World*, July 1919.

Essentially on the strength of his development work in radio during the war, and based on a law that protected unpatented "rights of industrial property" acquired during time of war, the Exchequer Court of Canada, on July 3, 1926, confirmed Grant's right to continue manufacturing broadcast receivers in Canada, freely using Armstrong's circuits. The Canadian Westinghouse Company appealed the decision to the Supreme Court of Canada, and on October 4, 1927, the decision was reversed. The judgment argued that the extent of the development work that Grant had conducted was insufficient to merit awarding him "rights of industrial property" to inventions that had been predominantly developed by Armstrong.

Grant never disputed the importance and the primacy of Armstrong's work. Rather, he felt that he had also made contributions to radio technology, and that under Canadian and British law he deserved access to the radio circuits that were by then in widespread use for commercial purposes after the war. Grant was very bitter about the Supreme Court decision. By then he was well established in the manufacture of broadcast receivers. He had models that used regenerative and reflex circuits, and by 1927 a superhet. Grant had to pull out of the receiver business, and returned exclusively to his practice as a broadcast engineer. He had to pay his own legal costs for the initial case, and all the costs of the appeal, and as a result had to sell his broadcast station CFCN in Calgary.

Grant was a charismatic character, and a good technical man. He had very little success in business, however, and seemed to be constantly in debt to his friends and family. He put tremendous energy into his technical work, but remained naive and subject to being taken advantage of in business matters. There was, for example, the matter of his one-time friend Gordon Love whose company sued him for patent violations. A year before the suit, Grant had personally sold him one of his new radios. As it later turned out, when Grant's broadcast station CFCN had to be sold to cover his debts, the buyer was Gordon Love, who had managed to remain successful in the broadcasting business during this period of turmoil for Grant.

Personal History

William W. Grant was born in Gibralter in 1892, where his father was a Warrant Officer in the Royal Engineers. In 1895 his father was posted to Halifax, Nova Scotia. Billie grew up in Halifax, attended public school there, and apparently tinkered with radio because when the war broke out, he enlisted in the 1st Canadian Divisional Signal Company. He was stationed in France throughout the war. In March of 1915, he transferred to the Royal Flying Corps, and in 1917 was promoted in the field to the rank of Second Lieutenant, Equipment Officer, Wireless.

After the armistice, Grant returned to Halifax. After several months he joined the Air Board of the

Canadian government. He became responsible for equipping search aircraft with radio, and with building base stations. In 1920 this job relocated him to High River, Alberta. For a time, he supplemented his government work with freelance activity—most notably, in 1921, in providing the technical capability for the Princess Theatre in Calgary to broadcast live concerts. At about the same time, he obtained a station license and built CFCN, a 50W broadcasting station in High River.

He was soon caught up in the commercial possibilities of broadcasting and, in 1922, resigned from the government radio service and moved his broadcasting station to Calgary. In the next several years, he increased the power of his broadcast transmitter (to 500W in 1924, to 10,000W in 1932), and built several other broadcast stations under contract (CKLC Red Deer, CJOC Lethbridge, CKUA Edmonton, CFQC Saskatoon). In 1924 he pioneered the use in Canada of phantom call letters, a device that allowed advertisers without broadcast licences to buy time slots on other stations. He also installed a monitoring station close to the U.S. border, and rebroadcast the U.S. shows while substituting their commercials with local ones.

He was not only a chain smoker of cigarettes but his chain smoking was developed to an interesting and imaginative degree. The cigarette was clamped permanently dead centre in his lips, so that the end of his nose was much yellowed from the smoke, while the two corners of his mouth remained free to alternately curse, and blow ashes from the cigarette at the person talking to him. At the same time he would use both hands to massage his eyeballs which continually stung from the smoke, thereby joggling his gold-rimmed glasses up and down over his face and forehead. He was described as "a man somewhat difficult to converse with." Being a chain smoker, he was not above appropriating his employees' cigarettes to keep a big enough supply on hand to last him through the twenty hour day he frequently worked, thus leaving his men smokeless through the long night hours when a cantankerous transmitter needed re-vamping for the next day's broadcast.

One visitor to a broadcasting station Grant was operating in 1928 observed that he had heated the transmitter shack by stringing haywire around the walls, nailed to the 2 by 4's, and energised by the 220 volt power line. Then, there was fresh oil spread on the floors to reduce the shock hazard from the electrified haywire. In the same shack he was melting the contents out of old power distribution transformers, rewinding them as modulation chokes, and melting fresh beeswax back into the cases. Do you suppose there could have been a fire? Well, there were three of them in his commercial premises, to the point where his insurers were getting nervous.

Grant never invested the energy and insight into his business affairs that he applied to his radio work. His income arrived in spurts and disappeared in spurts. He was fond of Packard cars, and had the capacity to spend lavishly even when he had little cash. After several years on his own as a radio station consultant, he joined the Canadian Broadcasting Corporation in 1935. In 1940 he obtained a leave of absence from the C.B.C. and enlisted in the Royal Canadian Air Force as a Pilot Officer. He spent the war mostly assigned to duties as a radio instructor, and was discharged with the rank of Squadron Leader. He then returned to the C.B.C., and later bought radio station CKLC in Kingston with his son Robert. He died in 1968.

Manufacturing

In the early days of W.W. Grant Radio Ltd., Northern Electric receivers were offered for sale. Until 1924, they were advertising the model R11 in combination with the R15 amplifier. In 1923 they also began the commercial manufacture of broadcast receivers. Eight models have been identified. The first sets were in mahogany boxes with hard rubber sloping front panels, and were called the "Grant Perfection" (see Figure 2). They consisted of three separate boxes strapped together: a tuner, a regenerative one tube detector and a two stage audio amplifier. Three UV 201 tubes were provided. A crystal set was also available at this time in the same style of cabinet.

The next year, a three tube set, called both "The Voice of the Prairie" and the "Grant Superdyne" was constructed in a single box with a vertical black

Fig. 2. The "Grant Perfection" Long Range Receiving Set

bakelite panel (see Figures 3 and 4). The Grant company engraved their own panels. Three R215-A peanut tubes protruded through the panel, allowing the rest of the components to be permanently enclosed in the box. The coils and tuning condensers were obtained from the New York Coil Co. Evidence of opening the box would void the warranty. No schematics were provided for W.W. Grant sets, but a diagram of this one set was provided as an exhibit in the Grant vs. Westinghouse trial (see Figure 5). This three tube set has a reflex circuit. The original price was $65.00 including tubes, phones, batteries and aerial equipment.

A four peanut tube version was introduced next, which was substantially the same set with the addition of another audio stage. In 1924 the company lost its inventory of bakelite panels in a fire. They had been stored in an unused elevator shaft. A four tube set was brought out again in 1927 with an etched brass panel (Figure 6). The panels were purchased from the Crowe Name Plate & Mfg. Co., Chicago. This set sold for $100.00 including all necessary equipment, and a Tower Meistersinger horn speaker. This was again a reflex circuit where the four tubes were advertised as providing two stages of r.f. amplification, a regenerative detector, and two stages of a.f. amplification.

There was also a five tube set, "The Voice of the Prairie Special". The early, bakelite panel version of this set used five peanut tubes, and appeared in a fine walnut cabinet. The price was $150.00 including tubes, batteries, phones and a Meistersinger horn speaker. The later version of this set (The Voice of the Prairie Five), had an etched brass front panel, and the final audio amplifier stage used a UX 120 tube.

Fig. 3. The "Grant Superdyne" Receiver

Fig. 4. Rear view of the "Grant Superdyne" panel

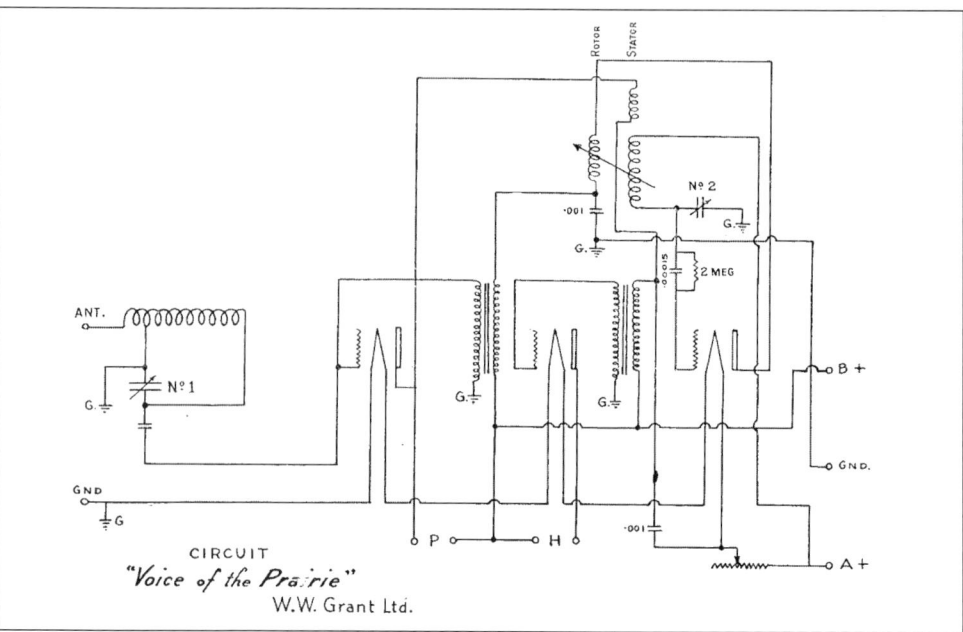

Fig. 5. Schematic diagram of the "Grant Superdyne" used in the court defense of the Canadian Westinghouse patent infringement suit.

Fig. 6. "The Voice of the Prairie Four" 1927 version with etched brass panel

Fig. 7. "The Voice of the Prairie Seven" superheterodyne receiver

Fig. 8. W.W. Grant crystal set with tuning fixed at 440M

Finally, a seven peanut tube superhet, "The Voice of the Prairie Seven", was introduced in an elaborate walnut cabinet (Figure 7). Shortly after its introduction, the Canadian Westinghouse Company won its appeal in the Supreme Court of Canada, and W.W. Grant Radio Ltd. ceased manufacturing broadcast receivers.

Again following the publication of this article in 1992, an Alberta collector gave me the crystal set in Figure 8, purportedly made by the W.W. Grant company. Its tuning was fixed at 440M, the wavelength of CFCN, his broadcasting station in Calgary. I am uncertain of the actual provenance of this receiver, having not seen another like it, or any other mention of one.

Acknowledgements

Thanks are due to Edgar Burge who provided his notes on this subject and a great deal of personal support. In the mid-1970's Ed interviewed a number of friends, family members and former employees of Grant, and has preserved a great deal of pertinent information upon which this article was largely based. Thanks are due to Bernard Payne who also provided information. The artefacts shown in Figures 3, 4, 6 and 8 are now in the Canada Science and Technology Museum, Ottawa.

REFERENCES

1. Canadian Westinghouse Co., Ltd. vs. W.W. Grant Limited et al., Exchequer Court of Canada. July 3, 1926, pp. 164-170.
2. Canadian Westinghouse Co. v. Grant et al., Supreme Court of Canada, October 4, 1927, Dominion Law Reports (1927), 4, pp. 484-494.

Fig. 9. WW Grant station CFCN studio, 1926

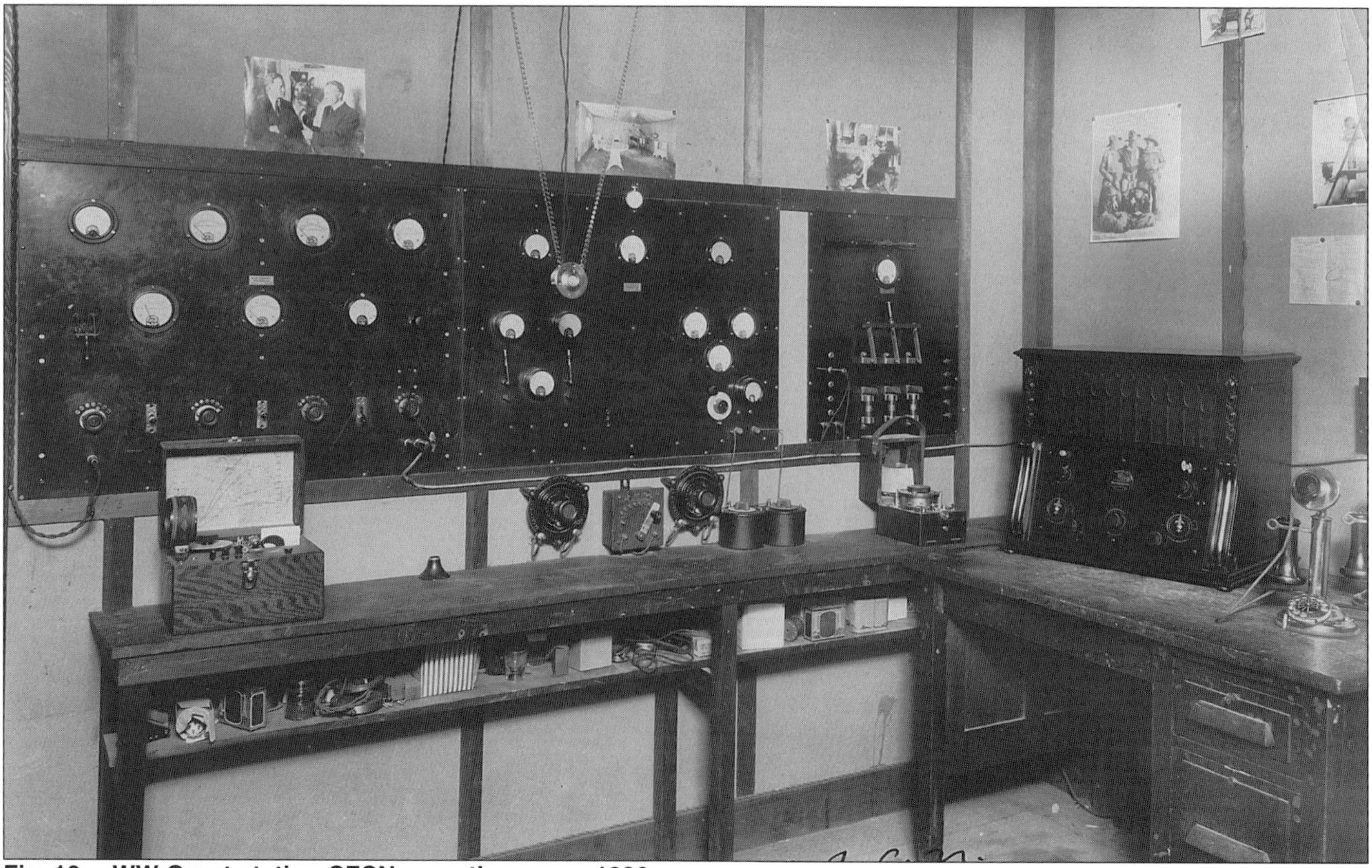

Fig. 10. WW Grant station CFCN operating room, 1926

~ 7 ~

Patent Avoidance – Kit Radios

The Eaton single circuit set is an example of how a company chose to provide kit radios rather than completed sets. The company of course happened to be a department store with a large mail-order catalog business.

Eaton Single Circuit Set, 1923
By Robert P. Murray

I never thought I would do it — dismantle a rare commercially made receiver of the 1920's and scatter the components into my parts bins. I assumed that after having been active in the hobby for ten years that I would be proficient in distinguishing commercially made from home made sets. This one turned out to be a special challenge.

It was designed by an engineer, Lynn V. Salton. It was a kit, so it was assembled by individuals with varying levels of skill. It was introduced at close to the low price end of the market, and accordingly used very inexpensive materials and techniques. Finally, the manufacturer's name appeared no where on it (Figures 1 and 2).

I knew of the existence of two of these sets. Besides mine, a collector friend Jerry Gamache had one. I dismantled mine but saved the parts, panel and box. Jerry saved his intact, except for the box which he discarded. We both discovered later what we had by seeing the ad in the Eaton's radio catalog (Figure 3). Eaton's was a national Canadian department store until 1999 when it was sold to Sears Canada. Jerry kindly lent me his receiver to help with my restoration.

The receiver used one type 215-A peanut tube. The instructions for assembly were almost nonexistent, but pre-formed bus wire (#14 unplated round copper) was supplied, and both a schematic diagram and a scale drawing. Extreme economies were taken in the physical design. The tube shelf consisted of a 3/8" wide strip of brass, bent as an L-bracket. It was fastened at the inside of the front panel to the rotating end of the rheostat shaft. The socket and tube

Fig. 1. The restored Eaton Single Circuit Set

were prevented from rotating as the rheostat knob was turned by their attachment to the stiff bus wire. The grid leak resistor was attached at one end to a small brass L-bracket extending from the bottom of the grid condenser. At the other end, a 3/16" brass eyelet was soldered to the bus wire to hold the resistor (Figure 4). The rheostat was an unusual type where the resistance wire fit in a groove around the back of the knob, and the wiper was fixed to the front of the panel.

The Eaton Single Circuit Set was introduced in 1923 as a kit including tube, headset, batteries, and aerial for the price of $29.75. The blueprint alone was priced at $.25. In the Spring of 1924 the price of the set was reduced to $28.00, and a two stage amplifier kit using peanut tubes was added for $22.50. In 1925 the receiver was offered for $21.95 and the amplifier

Fig. 2. The back of the panel in the Single Circuit Set

Fig. 4. Detail showing the tube shelf and the mounting of the grid leak resistor.

for $16.50. By 1926 the receiver kit was $19.95 and the amplifier was still $16.50. In the Fall catalog for 1926 the set was no longer offered. However, the Everyman crystal receiver was still available (the Canadian version of the de Forest). Its price, as a surplus item, was a mere $1.95!

The designer, Lynn V. Salton was born in 1897 in Moose Jaw, Saskatchewan, and educated in Winnipeg.[1] He studied arts at Wesley College in Winnipeg, and while in his third year in 1917 he enlisted in the British Navy as a wireless operator. In the course of the war he was promoted to a commissioned wireless officer. After the war he returned to Winnipeg to complete his degree, and graduated with the Gold Medal in Arts in 1920. He was then appointed Federal Radio Inspector for Western Canada, but resigned in 1922 in order to follow other business interests. He built the first commercial broadcasting station in Manitoba, for the *Winnipeg Free Press* newspaper in 1922. Its inaugural program, on April 2, consisted of a few musical numbers and a talk by the Reverend Dr. George F. Salton, Lynn's father.[2] Also in 1922 he was elected a member of the Institute of Radio Engineers, and on May 1, joined the T. Eaton Co. as their radio engineer. He remained with Eaton's for the duration of his career and eventually managed their radio retail business across Canada. In 1994, the Employee Relations group at Eaton's no longer had a record of his retirement date. One of his former employees from Winnipeg in the 1940's recalls him as, "a little peacock of a man, always very well dressed and strutting about."

Salton's Single Circuit Set came with little in the way of assembly instructions, and no operating instructions as far as I could tell. I was interested to find out how it would perform, and whether it would regenerate effectively. Being cynical by nature, I decided to examine the 1924 E.I. Co. publication *100 Radio Hook-ups*, to see whose design might have influenced Salton.[3] Except for the arrangement of the grid leak resistor, the circuit was identical to No. 28, The de Forest single coil Ultra-audion circuit. The author said,

"A single layer tapped coil is employed rather than a honeycomb coil, so that a less critical means of wavelength variation can be had. The Ultra-audion circuit, although very sensitive, is rather unstable in operation. Better control can be had by the use of a variable grid condenser having a maximum capacity of .0005 mfd."

I connected the receiver to antenna, ground and power source to hear it for myself. B+ was 25 volts obtained from a regulated ABC supply. The supply would not provide much under 2 volts at the A terminals, so I heated the filament of the 215-A with a "C" size alkaline cell. The antenna was a 70 ft. length of wire under the peak of my roof in an urban area, and the ground was a cold water pipe.

The results were mildly disappointing. The receiver covered the center of the broadcast band, from about 0.5 to 1.3 MHz with modest volume to

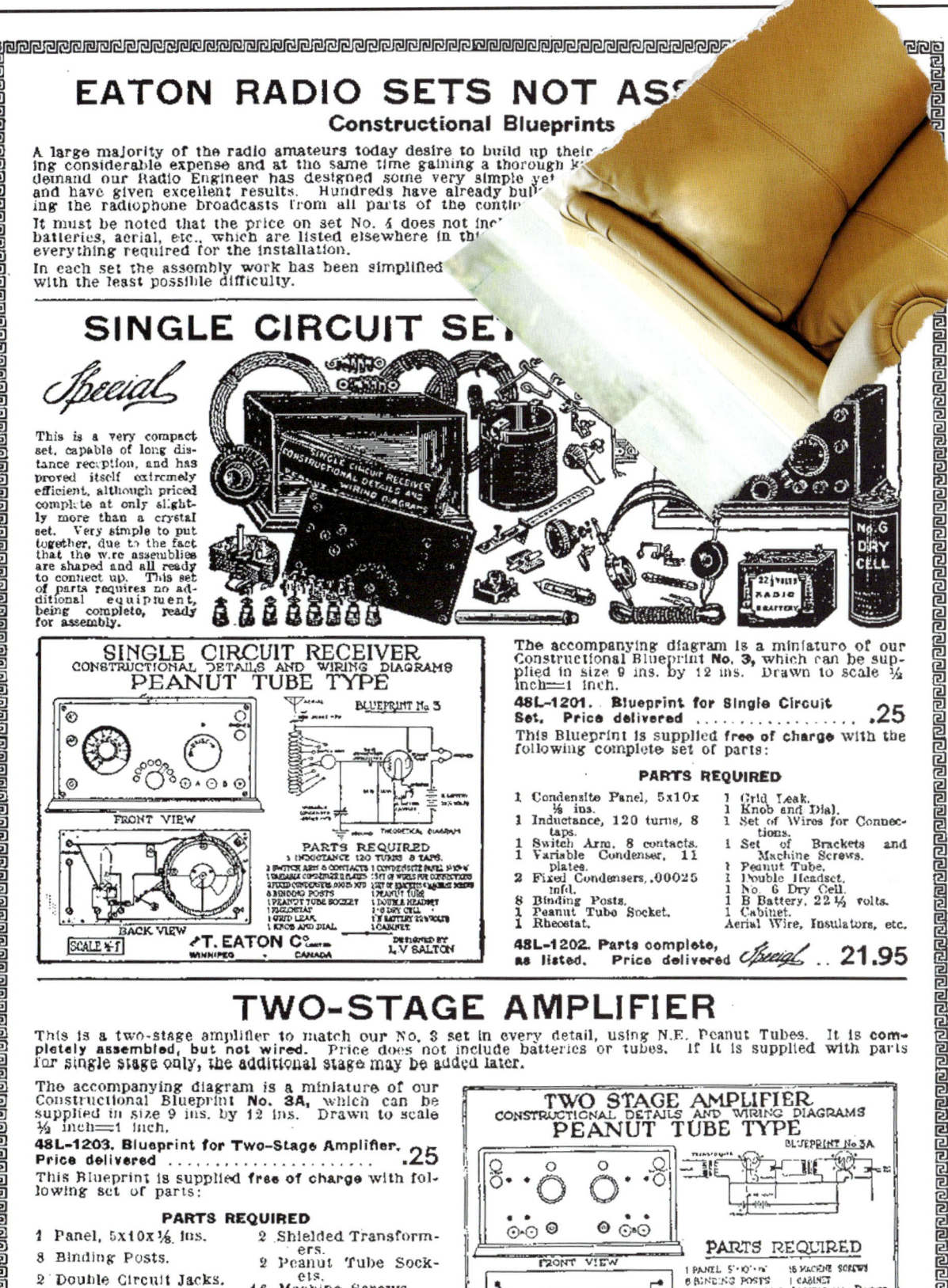

Fig. 3.   Page from the 1925 Eaton's Radio Catalog

the headphones. The tuning came predominantly from the eight point tap switch on the coil. The variable condenser had comparatively little effect. The net result was somewhat more selective and louder than commercial crystal sets I have tested recently, but not by much of a margin. Volume could be increased by adjusting the filament voltage, but I was not comfortable raising it much above 1.2 volts on the 215-A tube. None of the controls would throw the receiver into oscillation. With these results, I was not sure whether my reception was benefiting from regeneration. Regeneration was not mentioned anywhere in the Eaton's description, but the circuit was labelled regenerative in the E.I. Co. book. For considerable extra expense, the user had obtained a receiver that performed marginally better than a crystal set, but was subject to fading, not a crystal set characteristic.

Photo credit

The set in Figure 1 is from the author's collection, now in the Canada Science and Technology Museum.

REFERENCES

1. Maxwell, J.K. Prominent in Radio in Canada: Lynn V. Salton, B.A., M.I.R.E. *The Radio Bug*, 1923, 1, 11.
2. Reynolds, G.F. Early Wireless and Radio in Manitoba, 1909-1924. *History and Scientific Society of Manitoba Transactions*, 1979, Series 3, No. 35, 89-113.
3. Muhleman, M.L. *100 Radio Hook-ups*. New York: The E. I. Company, Second Edition, Revised, 1924.

~ 8 ~

The Mercury Super Ten

The Mercury was made by a motorcycle sales and service company in Toronto. The designer was C.A. Lowry and the circuit was a superheterodyne. H.M. Kipp did not have rights to use superheterodyne patents in Canada but reached some informal arrangement with Canadian Radio Patents Limited under which H.M. Kipp agreed not to publish their schematic. RCA at the time was preparing to face anti-trust charges in the U.S. and chose not to defend its superheterodyne patents aggressively. H.M. Kipp remained a small player. The Super Ten was designed to work in uniquely Canadian conditions of radio reception as they existed in 1925. (see Ammon, R.T. How a judge "unlocked" the superheterodyne circuit", *The Old Timer's Bulletin*, 2004; vol. 45: pp. 55-57.)

The Mercury Super Ten
By Robert P. Murray

The Mercury Super Ten was made in Toronto, beginning in 1925, by the H.M. Kipp Company. H.M. Kipp was primarily a motorcycle dealer but had some shop facilities and some established market presence. The receivers were advertised nationally from around 1925 and sold across Canada. They were especially popular in rural areas because their advertising emphasized their capability of receiving distant stations. The receivers were designed by C.A. Lowry, a young engineering student from the University of Toronto.

The receiver design was a superheterodyne with an intermediate frequency of around 247 kHz, not an optimum choice by today's standards but typical for its day. It was characterized by the use of ten type 215-A "peanut" tubes, the relatively reliable telephone industry repeater tubes. It had available plug-in coil sets covering 9-22 meters, 22-32 meters, 32-80 meters, 80-190 meters, 190-550 meters, and coils covering wavelengths up to 1,000 meters were available by special order.

The set was available either assembled or in kit form. Among the assembled sets, the DeLuxe model

Fig. 1. Mercury DeLuxe model, mahogany cabinet

came with vernier tuning and oscillator capacitors and hardwood (mahogany or walnut) cabinets. Two of these sets are shown in Figures 1 and 2, and there were other cabinets. The others were either supplied by H.M. Kipp or made elsewhere. H.M. Kipp was unlikely to have made the wooden cabinets in-house and a diversity of styles have survived. The radio was also available without a cabinet.

Five intermediate frequency (I.F.) transformers

were wound with cloth covered wire, probably silk covered, on wooden bobbins. Individual coils were insulated by squares of corrugated cardboard and the assembly was contained inside a rectangular copper box which was soldered closed (see Figure 3). The I.F. stages were peaked, all at the same time, by tightening or loosening two screws which protruded through the box and squeezed it. By todays standards, this was clearly not a finely tuned superhet. Battery power was connected by means of a cable ended with a Jones plug at the back of the set.

A lower-priced set, described simply as the Mercury Super Ten in the model "H" cabinet was available without vernier tuning and without plug-in coils (see Figures 4 and 5). The cabinet appeared to be made of pine and was described as having a walnut finish.

A later model of set with a cleaner arrangement of parts on the chassis became available in 1928 (see Figures 6 and 7). The front panel had a wooden border enclosing the controls, but still fit in the existing array of Mercury cabinets. The tuning and oscillator controls worked through vernier gears, and capacitors were located below the chassis with the I.F. coils and audio transformers (Figure 8). Only the plug-in coils and peanut tubes were located above the chassis. The chassis of the earlier sets were made of wood, and of the newer sets were made of bakelite. The newer sets had no distinguishing model number and as far as we know did not represent any circuit improvements.

Mercury Super Ten receivers turned out to be particularly well suited to the distribution of the population in Canada. In 1925 there were 60 broadcast stations in Canada compared to 700 in the United States. A Canadian listener wanting program variety had to reach out farther for his or her entertainment, often over 1,000 miles. Mercury receivers performed well over such distances, or perhaps just as important, they were believed to perform well. The Mercury Super Ten was made in Toronto, but today among collectors they are more commonly found in the prairie provinces and British Columbia.

Photo credits
All of the pictured sets are in the author's collection.

Fig. 2. Mercury DeLuxe model, walnut cabinet

Fig. 3. Mercury chassis showing the enclosed I.F. coils under the tube shelf, and plug-in R.F. and oscillator coils at either end.

Fig. 4. Mercury standard model without vernier dials

Fig. 5. Mercury standard model without plug-in coils

Fig. 6. Mercury 1928 model, front view

Fig. 7. Mercury 1928 model, top view

Fig. 8. Mercury 1928 model, bottom view

Before and After the Mercury Super Ten
Robert P. Murray

The Mercury Super Ten is a well known superhet of the 1920's. It was well designed for its time, using plug in coils for band changes, the economical and reliable 215A "peanut tube" (ten of them), and vernier tuning dials as an option with the more expensive models.[1-4] It had some disadvantages too, such as the lack of provision for peaking each i.f. stage separately, and an unfortunately low i.f. frequency at around 250 kHz, which was typical of the times. The engineer was Charles A. Lowry (see Fig. 1). He was an engineering student at the Univerity of Toronto from 1919 to 1923 but did not graduate.[5] He began work on the Super Ten in 1921, and in 1923 he joined the H.M. Kipp Company, which had been engaged in the business of selling and repairing bicycles and motorcycles. They were willing to manufacture and market his receiver, and they had a national sales force in Canada that was willing to diversify into radio.[6]

Before the Mercury

Mr. Lowry was the chief engineer and one of the officers of the Radio Devices Limited of Gananoque, Ontario, Canada during his engineering student days prior to 1923.[5] Mr. Lowry was presumably the company's engineer when the Model K crystal set was designed (see Fig. 2, 3). At least, for want of better information, we are tempted to blame him for it. The box is made from mahogany, but has no ornamentation. A very functional drawer handle adorns the lid. The panel is made from sheet aluminum, etched with dial scales and the company name. The aluminum panel was grounded, and probably provided some shielding that assisted with tuning the receiver (see Fig. 4).

The under side of the panel is where the surprise awaits us (see Fig. 5). The wiring of the set consists of aluminum straps fastened to machine screws with nickel plated nuts. I wondered at first if operators of this set had to dismantle it every month or so to clean the oxide from the junctions in order to make it work. I discovered that my specimen, not cleaned for the five years or so since I obtained it, worked quite well. Well, it sort of worked. With the 70 ft. antenna under the roof of my house, and a water pipe ground, it brought in local stations all across the broadcast band with a healthy volume to the headphones. The primary slider was set for the strongest signal level,

Fig. 1.   Charles A. Lowry

and the secondary slider tuned across the contemporary broadcast band. Unfortunately, it typically received around three stations at once. Further, the aluminum panel is springy, while the sliders are quite stiff to adjust. The result is that any movement of the sliders almost always knocked the cat's whisker off its point, so that the tuning had to be done in silence, and then the point on the crystal found again.

H.M. Kipp after the Mercury

The H.M. Kipp Company continued to manufacture Mr. Lowry's Mercury Super Ten with variations for about five years. I have had nine of them and no two were quite alike. Further, I have seen quite a number of minor variations different from the ones I have had. Lowry recalled that there had been four main models.[6] The first, in 1923, can be recognized by tuning condensers enclosed in cylinders of transparent celluloid. The second had a split-stator tuning condensor which enabled easier band switching by the use of different pins in the different plug in coils. No band switch as such was necessary. The third model, in 1925, had the condensors ganged to a single control. There was finally a model with AC tubes marketed after Lowry left in 1927, but with limited success.

A variation about which Lowry seems to have forgotten, or perhaps of which he was unaware after his departure, was the Mercury Super Power Ten. This was advertised in the 1929-30 catalog, and was the same superhet tubed with eight 201-C's and either 112-A or 171-A push pull output tubes. This configuration was offered either as a battery set, or with a power adapter as an electric set.

The company had continued to offer the Mercury to a point in time where it was clearly no longer a current design. More important was the company's loss of Mr. Lowry in 1927, over a dispute in which they had at first refused to convert the Mercury to use the new McCullough AC tubes manufactured by Rogers in Canada.[6] In total, more than 7,000 Mercury sets were sold.

It is interesting to speculate whether the H.M. Kipp Company may have ceased developments on the Mercury because they were using patents for which they were not licensed, as was the case with W.W. Grant in Calgary.[7] Lowry reports that they were indeed sued for patent infringement by the patent pool formed in 1926 under the name of Canadian Radio Patents Ltd., representing Canadian Westinghouse, Northern Electric, Canadian General Electric, Canadian Marconi and Rogers.[6] He recalls that in his examination for discovery, he reported that at least four of the patents cited against the H.M. Kipp Company were in dispute elsewhere and would risk being revoked if the case came to court in Canada. The suit was withdrawn and H.M. Kipp Company were allowed by the patent pool to continue to operate, under the condition that the circuit schematics of the receiver would not be published. The schematic was subsequently published in the service manuals of the Radio College of Canada, but presumably after production had ceased.

The next offering of the H.M. Kipp Company was what we now refer to as a cathedral, and was labelled the Mercury Mantel Set in the Radio College of Canada manuals, and the Mercury Jr. in newspaper advertising (Fig. 6, 7). It was in a wooden cabinet of not very charming design. Furthermore, this manufacturer with six years of experience in the production of superhets was now offering a TRF. It was an electric set of undistinguished design. My specimen had been submitted to a couple of highly intrusive repairs, but still it was evident that the original workmanship had been barely adequate.

Another strategy tried by the H.M. Kipp Company in 1931 was to market a ready made receiver, the Echophone S5, likely suggesting that their own cathedral had not been a great success. On the back of the S5, coexisting with the Echophone labelling indicating manufacture in Waukegan, Illinois, the Kipp Company used their remaining supply of identification plates. My S5 therefore also states that it was made in Toronto by the H.M. Kipp Company Ltd. Oh well, at least it's a superhet.

At the end of 1931, the H.M. Kipp Company was

Fig. 2. Radio Devices Limited, Model K case

Fig. 3. Model K crystal set

soliciting dealer franchises in a radio trade magazine, and advertising a series of console models, including one twelve tube model with short wave as well as standard broadcast coverage. (Well, eleven tubes and a 'voltage regulator'.) Tube types 224(1), 227(3), 235(4), 247(2) and 280(1) were used. In the same issue of the magazine was an article by Herbert W. Kipp (son of the founder), warning that customers would find the reception of short wave broadcasts a challenging and unreliable pursuit.[8] The 1931-32 models were not shown in the Radio College of Canada manuals, and apparently were the last of the company's radio offerings.

Charles A. Lowry after the Mercury

Mr. Lowry went on to have a distinguished and varied career as an engineer. From his resume, the following is a summary:[5]

• 1927-33 — Plant radio engineer, de Forest Radio Corporation of Canada.

• 1933-35 — Chief engineer, Polymet/Aerovox of Canada

• 1935-39 — Product engineer, Stromberg-Carlson,

Fig. 4. Model K crystal set aluminum panel

Fig. 5. Bottom of crystal set panel

Fig. 6. Mercury Mantel Set, 1931

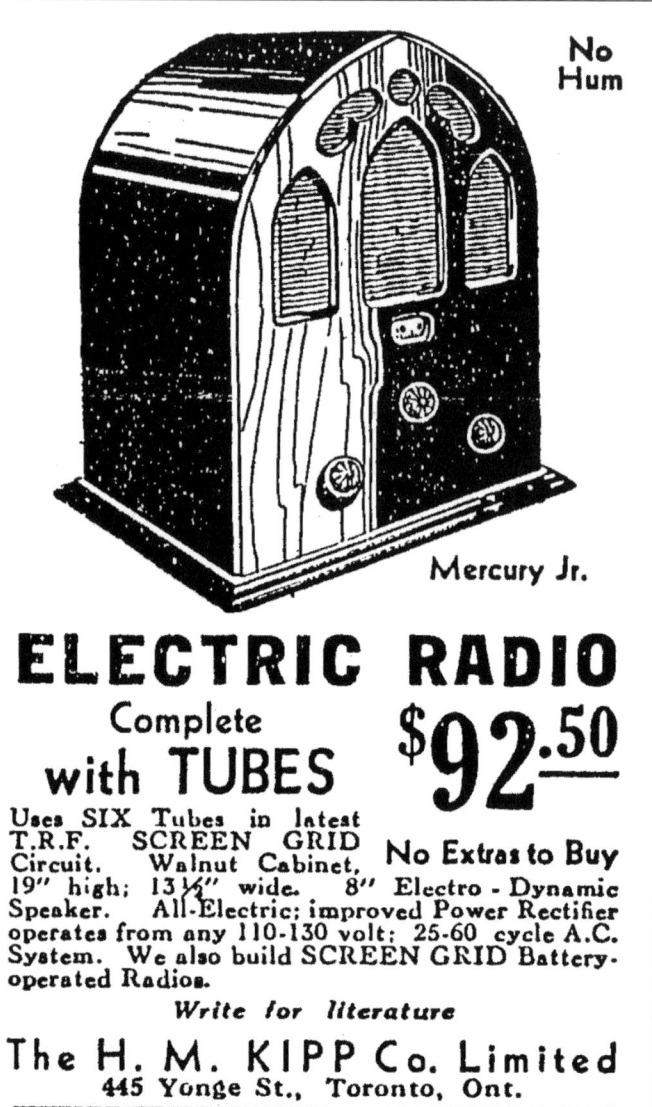

Fig. 7. Ad for Mercury Jr. 1931

Rochester, NY
- 1939-40 — Production engineer, Stromberg-Carlson (Australasia), Sydney, N.S.W.
- 1941-42 — Engineer, Syme, E.S.M. Pty. Ltd., Sydney, Australia.
- 1942-44 — Project engineer, Commonwealth Aircraft Corporation, Sydney.
- 1944-45 — Engineer, Philips Electrical Industries of Australia Pty. Ltd., Sydney.
- 1945-51 — Examiner of patents, Commonwealth Patent Office, Canberra, A.C.T., Australia.
- 1952 — Patent writer and consultant, Marsden and Bromley, Patent Attorneys, Toronto, Canada.
- 1953-1971 — Advanced planning engineer, Philco Corporation of Canada, Ltd., Toronto.

Charles Lowry retired from Philco at age 76. One of his last projects there was as systems engineer and project coordinator for the development of automatic alignment testing for auto radio production. The following is a memo written by Gerry Bruun, Manager of Engineering:[5]

> "With the completion of the above project, it would seem appropriate to acknowledge in a special way the major contribution which was made by the senior member of the engineering team. Mr. C. Lowry had been assigned to this very complex project as the advance planning and the electro-mechanical systems engineer, for a period of about 1-3/4 years (1968/69). Mr. Lowry designed nearly all the mechanical parts and assembly layouts for this equipment. Moreover, he physically produced a great many of these parts in the shop and he almost single-handedly assembled the original prototype. In view of his age (74 years) this was an exceptional achievement, which he accomplished with such remarkable vigor and efficiency that a younger man would be hard put to do better."

Throughout his long career Lowry had apparently given less attention to profit considerations in favor of his striving to solve technical problems. Sometimes, if his sense of the technical requirements of a situation clashed with the ideas of those around him, he would move elsewhere.[9] Charles Lowry remained an active radio amateur until he died in 1976 at the age of 81. To the end he was highly regarded as an innovative engineer, and a dedicated friend to those close to him.

Acknowledgements

My thanks to Dr. Charlie Fisher, enthusiast of early Canadian superhets, for his comments on this article, and to Bob MacIntyre and Roy Sawley who provided helpful information. Special thanks to my finder of obscure artifacts of early Canadian radio history, who shall remain anonymous. The Radio Devices Limited Model K crystal set described here is now in the Canada Science and Technology Museum.

REFERENCES

1. Radio Informer. The story of the Mercury Super-Ten. *Radio Informer*, October, 1925.
2. Allen, R. The Mercury Super Ten. *The Old Timer's Bulletin*, 1985: 26(3); 18-19.
3. Patterson, D. The story of Mercury Super 10. *Radio Age*, 1990: 15(5); 1-3.
4. Rhodes, C. Project Mercury: The restoration of a Mercury superhet. *Radio Age*, 1990: 15(5); 5-7.
5. Millen, T.I. Eulogy: Charles Lowry. *The Cat's Whisker*, 1977: 7(1); 4-6.
6. Lowry, C.A. Letter to Mike Batsch Re: Mercury Super Ten. *The Cat's Whisker*, 1976: 6(3); 4-8.
7. Murray, R.P. "The Voice of the Prairie": A brief history of W.W. Grant (1892-1968). *The Old Timer's Bulletin*, 1992: 33(3); 16-19, and Chapter 6 in this book.
8. Kipp, H.W. Pioneer short wave manufacturer says dealers should not promise too much. *Radio Sales*, 1931: 8(12); 12.
9. Challoner, D. Editorial: Charles A. Lowry. *The Cat's Whisker*, 1977: 7(1); 3.

Yuletide Entertainment

brought in by TEN –
Northern Electric Peanut Tubes

> Richdale, Alta.
> April 12th, 1926.
>
> The H. M. Kipp Co., Limited,
> 447 Yonge St.,
> Toronto, Ont.
>
> Dear Sirs:—
>
> I received the MERCURY SUPER TEN on the 23rd of March, and must say that I am more than delighted with it. I do not think that you sound half its praises in your advertising matter.
>
> I can speak from experience, when I say that the set will do all that you say that it will, and more.
>
> All who have seen the set admire it very much, both for its wonderful compactness and size. It seems incredible that so much power can emanate from the set. Our piano and cabinet grand gramophone are having quite a rest now since we can get such wonderful results from the air through the means of the MERCURY SUPER TEN.
>
> I am never tired of sounding its praises to the many people that have already called in to see it; and if I can put any business your way I will be only too glad to do so.
>
> Again thanking you,
> Yours most sincerely,
> "J. G. C."
>
> (Name on request)

'Tis Christmas Eve. The firelight casts a soft glow over the quiet room. Then voices—silvery voices from afar, joining in sweet harmony in the old carols—'Hark the Herald Angels Sing,' 'Merry Merry, Merry Christmas Bells,' 'Peace on Earth,'' and the ringing tones of a male chorus as they sing: 'God Bless you, Merry Gentlemen,' .... Wonderful Yuletide entertainment, faithfully reproduced by the MERCURY SUPER-TEN.

The *Mercury* with its *Ten* Peanut Tubes will operate any loud speaker without additional amplifiers. It is as selective as a telephone line and cuts out local stations in a degree or two on the dials. The Mercury is a wonderful loop aerial receiver and due to the directional qualities of a loop, interference from stations operating on approximately the same wave lengths is materially reduced. The Mercury stands in a class by itself for long distance reception and is the winner of the Silver Trophy for Summer Reception. Despite its giant power the Mercury is economical in current consumption. The *Ten* Peanut Tubes use less battery current than *Two* ordinary audion bulbs.

Before you purchase a set at any price write for our free 24 page booklet, "The Story of the Mercury Super-Ten."

A WONDERFUL OPPORTUNITY FOR LIVE DEALERS

THE H. M. KIPP COMPANY, LIMITED
447 Yonge St.   Toronto

~ 9 ~
The Component Parts of the Radio Corporation of America

RCA was formed initially in 1919 from the American Marconi Company and the General Electric Company. The Westinghouse company soon joined in. The Great War of 1914-18 was fresh in people's minds, and the U.S. government feared the possibility of another such situation where wireless communication would be in foreign hands — controlled by England.

In Canada the prevailing sentiment at the time was more in sympathy with actions by England — she was the mother country after all. There was no parallel attempt to dislodge the Canadian Marconi Company.

Manufacturing of receivers by RCA was placed in the hands of General Electric and Westinghouse. The manufacturing and sales by the subsidiaries in Canada, Canadian General Electric and Canadian Westinghouse, casts an interesting light on the activities at RCA, and that is why the following article was first written. It also provides an interesting reflection on the subsidiaries' activities in Canada and that is why it is included here.

Some of the non-RCA radio design and marketing in Canada, at least by one of the subsidiaries CGE, is further illuminated by one of their Canadian engineers, C.L. Richardson.

Broadcast Receiver Manufacture by General Electric and Westinghouse in the First Decade of RCA
By Robert P. Murray

This story begins with the formation of the Radio Corporation of America (RCA) in October, 1919 from the American Marconi Company. General Electric (GE) and the U.S. Navy were key players.

Radio was, in 1919, a point-to-point communications medium, although that would change a scant few years later. American Marconi's parent company in England had fallen behind in technology — the company still relied on spark transmitters in 1919. General Electric had the Alexanderson alternator, the best performing transmitter of the day. American Marconi had been negotiating with GE to obtain exclusive world rights to the alternator technology, but the Navy was reluctant to allow it. During World War I the U.S. Navy had taken control of all U.S.-based communications, and for a time the Navy had hoped they might continue that control in peacetime. Had American Marconi been able to finalize a deal with GE for the rights to the Alexanderson alternator, it would have created a situation, as it then appeared from the American perspective, where effective control of radio transmission would fall into foreign hands.

The fear that a non-U.S. entity (Marconi) would control overseas radio communications, a technology deemed critical to U.S. national security, was so strong and so widely held that the American government (primarily the U.S. Navy Department), was prepared to allow a monopoly within the radio industry if that was the only way to keep control of radio in U.S. hands. President Woodrow Wilson, attending the 1919 Paris Peace Conference, accelerated the changes by sending a message to the director of naval communications "that he counted on him to keep a careful watch on American interests in radio" (Aitkin, 1985, p. 280). This statement was later amplified to mean that American national interests were at stake.

The negotiators for American radio interests at the time feared British domination, but in so doing overestimated the strength of the British economy after the war (Aitken, 1985). It has been suggested that the American Marconi Company was expropriated by the U.S. Government (Herron, 1969 for example), but most historians do not describe it that way.

Rather, it was more of a friendly corporate takeover with the strong encouragement of the Navy. Aitken describes the negotiation process in painstaking detail, and suggests that GE eventually paid somewhat more than the market price for American Marconi stock. On the other hand, it was widely recognized that there was a strong political will to create an all-American radio company after World War I, and this likely inflated the perceived value of American Marconi stock in the transaction. In the end, both parties seemed reasonably satisfied with the price per share.

The Navy had urged GE to act quickly. The initial moves were by Owen Young, GE's Vice President, and Edward J. Nally, American Marconi's, Vice President, who met and concluded that a merger would benefit them both. Under the arrangement, American Marconi (to become RCA) would be the operating company and GE would focus on manufacturing. As the plan solidified, the notion of a government radio monopoly under control of the Navy faded (due apparently to effective congressional lobbying by David Sarnoff, Nally's assistant at the time), and the idea of a private monopoly took its place. In fact there was widespread dissatisfaction with the way the government had run the public utilities during the War. Public interest in a government-run radio monopoly was, no doubt, overestimated by the navy (S.J. Douglas, 1987).

The American Telephone & Telegraph Company, and its subsidiary the Western Electric Company joined RCA in 1920. The Westinghouse Electric & Manufacturing Company, the United Fruit Company and its subsidiary the Wireless Specialty Apparatus Company joined in 1921. A recent account of events leading up to the formation of RCA is provided by Lewis (1991a). S.J. Douglas (1987) and Howeth (1963) also provide helpful information regarding the formation of RCA.

An earlier version written by an academic closer in time to the events and with the blessings of Sarnoff, by then president of RCA, was the historical volume *Big Business and Radio* (Archer, 1939). Archer emphasized that the agreement between the corporate entities, signed from 1919 to 1921, contained two main matters of importance. First, the participants each agreed to share in access to the others' patents. This created a patent environment similar to that to which the companies had become accustomed during the War, and greatly facilitated the design and manufacture of equipment using the best known techniques of the time. The second important content of the agreement was the appointment of RCA as the sales outlet of the participating corporations. Structured this way, RCA could overcome the earlier competitive attitudes of its corporate partners, or at least partially overcome them as we shall see.

Archer (1939) goes on to say that if radio had turned out to be only the passing of commercial traffic from point to point as had been RCA's original purpose, the company would have had trouble surviving. What occurred next, however, was referred to as the "Radio Boom", and RCA was initially unprepared for it.

Archer wrote a relatively non-critical view of the industry, with the overt cooperation of the RCA president. Consequently, his book is highly sympathetic to RCA. Let me digress to yet another contemporary view. As rosy a picture as Sarnoff is credited as taking of his role in the development of RCA, Gutterman (1968) has outdone him. Sarnoff cooperated in the creation of this biography, which states, for example, "World renowned as 'The Father of Radio and Television,' Sarnoff stands by himself among the great American men of wisdom. ... In the Radio Corporation of America, David Sarnoff created out of many diverse units one great corporate institution covering the entire field of electronics communication, from research through production. ... Guiding genius of a great organization, no other business leader has done more for the enlightenment and independence of American thought through radio broadcasting and television. His life is a tribute to the dynamics of freedom." (pp. 9,10). Gutterman goes on to include testimonials from 34 eminent Americans, presumably largely or entirely fiction. To his credit, Sarnoff found the exaggerated praise in the book offensive, and after seeing it tried vigorously to recall the sold copies.

Still, it was Sarnoff's position that the organizational strategy of RCA was important to present to the world as largely a product of his own foresight, even if he had to revise history somewhat to have the record say so (Lewis, 1991a; 1991b). There remains a possibility, therefore, that if in the conceived organization some aspects did not go entirely according to plan, we may not have heard of it. Certainly, Archer and Gutterman did not dwell on RCA's failures.

Manufacturing for RCA

In the 1919 agreement, GE was responsible for manufacturing equipment sold by RCA. (What little manufacturing capability American Marconi had was absorbed by GE.) When Westinghouse joined the group, responsibility for manufacturing was reallocated, with 60% to GE and 40% to Westinghouse. By

1921 broadcasting to the public, and the resulting demand for receivers and transmitters, was increasingly being recognized as a possible business opportunity. Among the main partners in RCA, Western Electric was to manufacture broadcast transmitters, and GE and Westinghouse were to manufacture receivers. Wireless Specialty Apparatus (WSA) made only a negligible number of receivers.

Anderson (1990) describes some of the early production at GE. Archer (1939) details the early broadcast receiver production of these two partners (Table 1). The plan was then to have each of these companies submit designs of new receivers to a Receiver Manufacturing and Design Committee at RCA for approval. If approved, designs were to be manufactured by both GE and Westinghouse to the same specifications — RCA specifications. Archer reports that RCA had initial difficulty in coordinating manufacturing standards between GE and Westinghouse, a factor in the relative advantage found by the independent manufacturers of the time. Competitors such as Atwater Kent, Crosley, Grebe, Grigsby-Grunow, Freed-Eisemann, Fada, Philco, Zenith and others got a head start in the home broadcast marketplace while RCA was working out its internal policies.

The requirement that engineers from GE and Westinghouse were to submit plans to a Receiver Manufacturing and Design Committee for approval apparently caused considerable turmoil. Barnum (1991) reports that "RCA faced tremendous coordination problems as a result of the dispersed competing operations" (p. 142). Alfred Goldsmith was in charge of the RCA Technical and Test Department at Van Cortlandt Park in New York City. Receiver prototypes were submitted independently by GE and Westinghouse. They were tested and the results were reported back by the Receiver Manufacturing and Design Committee. The approved models were then to be built by both companies. "However, due to diverse opinions among the various representatives of GE, Westinghouse and RCA, agreements were difficult to reach ... As a result, progress in product development and manufacturing was slow for the next several years" (Barnum, 1991, p. 143).

Under these circumstances it is not surprising that most of the histories of RCA and its players do not mention the relative difficulty of the dual manufacturing of receivers. The purpose of this paper to estimate whether broadcast receivers in the first decade of RCA were in general manufactured separately by General Electric and Westinghouse to common specifications, or whether there was a residual corporate rivalry that resulted in some models being made by one company and some by the other.

Methods

Several approaches were used to trace the provenance of receivers. First was the direct mention in credible published sources such as Archer, *Big Business and Radio* (1939), RCA, *Radio Enters the Home* (1922), and Barnum, *"His Master's Voice" In America* (1991). Dr. John Anderson, retired from GE, undertook to search relevant material from the Schenectady, NY, Museum which houses the GE Archives. He suggested that broadcast receiver manuals that were printed in Schenectady would be an indicator that the set was manufactured there. He suggested also that receivers mentioned in the *Works News* (company newsletter) would likely have been manufactured by GE. Finally he went manually through the photo cards in the Schenectady Museum, suggesting that receivers with more photos would likely have been manufactured by GE, compared to receivers with no or fewer archival photos. He did not suggest an absolute cutoff number of photos that would indicate GE manufacture. Regrettably, no correspondingly valuable information has been recovered from Westinghouse.

Table 1
Company of Manufacture for Early RCA Sets
(Adapted from a 1923 price list)

Model	Manufacturer
Radiola I	GE
Radiola II	GE
Radiola IV	GE
Radiola V	GE
Radiola VI	GE
Radiola Senior	Westinghouse
AC two stage amplifier	Westinghouse
RS	Westinghouse
RG Radiola Grand	Westinghouse
RT Antenna coupler	Westinghouse
RA Tuner	Westinghouse
AR Radio amplifier	Westinghouse
DA Detector-amplifier	Westinghouse
AA 1520 Radio amplifier	GE
AA 485	Wireless Specialty

Adapted from Archer (1939), page 16.

Collecting early broadcast receivers in Canada gives one a slightly different perspective on the manufacturing efforts of GE and Westinghouse in comparison to each other. It occurred to several of us Canadian collectors that the subsidiaries of General Electric (CGE) and Westinghouse in Canada sold broadcast receivers in the 1920's, and RCA did not operate in Canada in that decade. Accordingly the subsidiaries, CGE and Canadian Westinghouse, each sold receivers labeled with their own corporate names.

Further we assumed that CGE would not be motivated to sell in Canada a receiver that had been manufactured by Westinghouse in the U.S., nor would Canadian Westinghouse be motivated to sell a receiver in Canada that had been made by GE in the U.S. We therefore inferred that the labeling of receivers in Canada would be another indication of the manufacture of receivers in the U.S. To further check on the connection between receivers sold in Canada and similar receivers made in the U.S. by the parent companies and sold under the RCA label, we checked whether service schematics in Riders manuals and in Radio College of Canada (RCC) manuals were identical for sets bearing the same model number.

A tabulation of all of these indicators of provenance of manufacture is shown in Table 2. The table attempts to cover all major models of receiver sold by RCA between 1921 and 1929. There were a few models made by the Canadian subsidiaries and not by the U.S. parent companies and these are also included. A few corporate idiosyncrasies complicate the picture. CGE, when it offers receivers in Canada matching those with RCA model numbers in the US, tends usually to offer a receiver that is identical to the RCA model. Canadian Westinghouse on the other hand, when offering receivers matching those with RCA model numbers, tends more often to provide receivers with design modifications compared to the RCA model.

Results

Early sets with Manufacturers Identified by RCA

As will be seen in Tables 1 and 2, there is a group of receivers manufactured in the years 1921-1923 whose company of manufacture is identified by Archer (1939, pp. 15, 16). Since Archer was writing less than twenty years after the events, and since he had generous access to RCA documents, it appears safe to assume that this information was accurate (Table 2, footnote 1). The RCA publication *Radio Enters the Home* also appeared before RCA saw itself as creating the impression externally that the corporate origins of receivers were not to be identified, so it indicates the respective manufacturers (Table 2, footnote 2). Barnum (1981), who worked largely with source documents in the Hagley Museum, Wilmington, Deleware, is another credible source who essentially agrees with the others on the origins of early sets (Table 2, footnote 3).

Information from the Schenectady Museum

Instruction manuals printed in Schenectady (Table 2, footnote 4) are taken to indicate manufacture by GE. Mention in the *Works News* at GE is also taken as an indicator of manufacture by GE (Table 2, footnote 5). The number of photo cards in the Schenectady Museum has been recorded by Anderson (2003). These are shown in parentheses beside the footnote reference 6 in Table 2. These numbers vary widely between models, and are taken to indicate manufacture by GE if greater than 2. No such criterion number was suggested by Anderson.

Information from the comparison of Riders and RCC manuals

Comparing schematics in Riders radio manuals produced in the U.S. with those in the RCC manuals produced in Canada (Footnote 7 of Table 2) provides additional information about which Canadian models are similar or identical to their U.S. counterparts. The seventh column in Table 2 indicates whether the corresponding schematics are the same.

Similarities and Differences

In the following pages we will consider some of the sets made in both the U.S. and in Canada, approximately in the order that they appear in Table 2 (numerically by model number).

The crystal set model ER-753, later called the Radiola I, was made by GE according to several early sources which all agree. CGE sold a crystal set model 118, and Canadian Westinghouse did not make a crystal set. If we set out to confirm that GE and CGE both made these sets (we don't really need to confirm it), we can compare the appearance of the Radiola I with that of the CGE 118. Consider the CGE set in Fig. 1 in comparison to the Radiola I as pictured in Douglas, 1991, volume 3, page 14 or Bunis & Bunis, 1992, second edition, page 157. There are a few common features but also prominent differences. The Radiola I has a perikon detector while the CGE 118 has a galena detector. They both have embossed black panels but the knob arrangement is different and the boxes are different. They have similar binding nuts and a similar tuning knob. In sum we have to concede that the comparison of these two sets does

Table 2
Index of Broadcast Receivers Sold by RCA between 1921 and 1929

USA manufacture				Canadian manufacture			Fig.
RCA Designation	Year	Footnotes	Made by	CGE	Cdn Westinghouse	Riders & RCC agree?	
RA	1921	1,2,3	West				
DA	1921	1,2,3	West				
RC	1922	1,2,3	West		RC		
Aeriola/Radiola RS	1923	1	West		Aeriola RS		
Radiola AR	1923		West				
AR 1300	1922	1,2,3	GE				
AR 1375	1922	1,2	WSA				
AA 1400	1922	1,2,3,6(1)	GE				
Radiola I (ER-753)	1922	1,2,3,5,6(2)	GE	Model 118			1
Aeriola Jr.	1921	1,2,3	West				
Aeriola/Radiola Sr.	1922	1,2,3	West		Aeriola Sr		2,3
Aeriola Grand	1922	1,2,3	West				
Aeriola/Radiola AC	1922	1	West		Radiola AC		
Radiola Sr. Amp.	1923	1	West				
Radiola Grand	1923	1,6(1)	West				
AA 1520	1922	1,6(1)	GE				
Radiola II	1923	1,3,4,6(17)	GE				
Radiola III	1924	3,6(2)	GE,West	Radiola III	Radiola III	unk/yes	4-9
Radiola IIIA	1924	6(2)	GE,West	Radiola IIIA	Radiola IIIA	unk/no	10,11
Radiola Bal. Amp.	1924		West		Radiola Bal. Amp	no	
Radiola IV	1922	1,4,6(36)	GE				
Radiola V	1922	1,6(7)	GE				
Radiola VI	1922	1,3,6(6)	GE				
Radiola VII	1923		GE				
Radiola VIIB	1924	6(2)					
Radiola Super. AR812	1924	4,6(12)	GE,West	Radiola Super. AR812	Radiola Super	unknown	12,13
Radiola Super VIII	1924	4,6(45)	GE	Radiola Super VIII	Model 8	no	14
Radiola IX	1923	6(1)	WSA				
Radiola X	1924	6(6)	West		Radiola X	unknown	15
Radiola Regenoflex	1924		West		Radiola Regenoflex	unknown	23
Radiola 16	1927	5,6(14),7	GE	Radiola 16	Model 16	yes/no	16,17
	1928				Battery Operated		18
Radiola 17	1927	6(26),7	GE	Radiola 17		yes	
Radiola 18	1928	6(20),7	GE	Radiola 18		yes	19,20
	1927				Batteryless Receiver		22
Radiola 19		6(15)	GE				
Radiola 20	1925	5,6(18),7	GE,West	Radiola 20	Radiola 20	yes/unkn.	
Radiola 21	1929	6(8),7	GE	Radiola 21		yes	
Radiola 22	1929	6(15),7	GE	Radiola 22		yes	
Radiola 24	1925	4,6(9)	GE	Radiola 24			
Radiola 25	1925	4,6(50)	GE	Radiola 25			

Table 2 (Continued)
Index of Broadcast Receivers Sold by RCA between 1921 and 1929

USA manufacture				Canadian manufacture			Fig.
RCA Designation	Year	Footnotes	Made by	CGE	Cdn Westinghouse	Riders & RCC agree?	
Radiola 26	1925	5	GE				
Radiola 28	1925	4,5,6(45),7	GE	Radiola 28		yes	
Radiola 29		6(8)					
Radiola 30	1925						
Radiola 30-A	1927	4,5,6(25)	GE	Radiola 30-A		no	
Radiola 32	1927	5,6(18)	GE	Radiola 32		no	
Radiola 33	1929	6(6),7	GE	Radiola 33		yes	
Radiola 33 DC	1929		GE				
Radiola 41	1928						
Radiola 44	1929	6(23),7	GE	Radiola 44		yes	
Radiola 46	1929	6(3),7	GE	Radiola 46		yes	
Radiola 47	1929	7	GE	Radiola 47		yes	
Radiola 48	1930						
Radiola 50	1927	6(4),7	GE	Radiola 50		yes	
Radiola 51	1928	7	GE	Radiola 51		yes	24,25
	1925				Model 53		26
Radiola 55	1928	6(5)			Model 55 (1925)	no	27
	1926				Model 56		
	1926				Model 57 Special		28
	1927				Model 58		
Radiola 60	1928	6(2),7	GE	Radiola 60	Model 60	yes/no	29
Radiola 62	1928	7	GE	Radiola 62		yes	
Radiola 64	1928	7	GE	Radiola 64		yes	
Radiola 66	1929	6(4),7	GE	Radiola 66		yes	
Radiola 67	1929	7	GE	Radiola 67		yes	
	1929				Model 69		
	1929				Model 89		
	1925				Model 93, 193		30
	1929				Model 99		
	1925			Model 205			

FOOTNOTES:

1. Archer (1939)
2. RCA (1922)
3. Barnum (1991)
4. Anderson (2003) manuals printed in Schenectady
5. Anderson (2003) mention in *Works News*
6. Anderson (2003) no. of photo cards in Schenectady museum
7. Schematics in *Riders* and *Radio College of Canada* manuals are identical.

not advance our argument for similar manufacture from General Electric and CGE very well.

The Aeriola Sr. is one of the early receivers that was identified as RCA in 1921, but also labeled as made by Westinghouse (Fig. 2). The Westinghouse subsidiary in Canada offered an identical set, except for labels that indicated manufacture by Canadian Westinghouse in Hamilton, Ontario (Fig. 3). The location of these labels on the Canadian set was in the paper lid instruction card, and on the tuning dial scaled 1 to 10. In fact the replacement dial scale was on an aluminum plate fixed on top of the original, which still indicated Westinghouse's East Springfield Works.

A confusing comparison is around the production of Radiola III in 1924. The Canadian Westinghouse, RCA and CGE versions are shown in Figs. 4, 5 and 6 respectively. They look quite similar externally, although the CGE model is finished in gold wash, rather than the nickel plating of the RCA and Canadian Westinghouse models. Also, both RCA and Canadian Westinghouse models used WD-11 tubes, while the CGE used UX-199 tubes. The undersides of the chassis are shown in Figs. 7, 8 and 9 respectively for Canadian Westinghouse, RCA and CGE. Again, the Canadian Westinghouse and RCA models are visibly similar using identically-shaped closed transformers and a row of cylindrical capacitors along the right

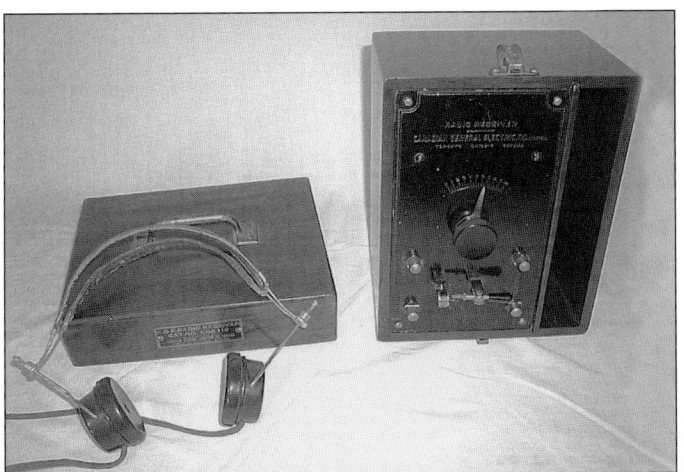

Fig. 1. CGE model 118 crystal set (Robert MacIntyre collection)

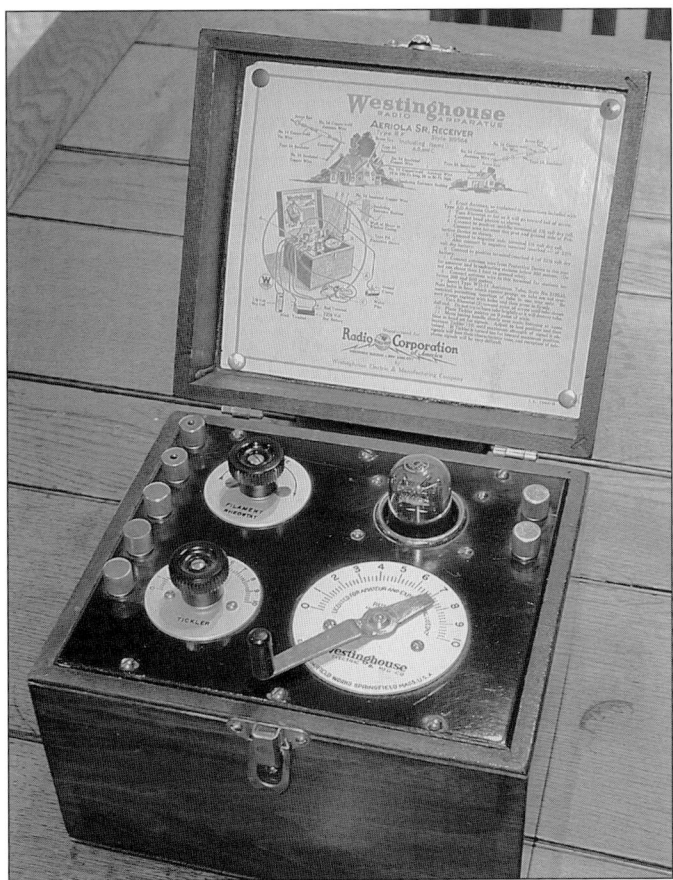

Fig. 3. Westinghouse Aereola Sr. (Author's collection)

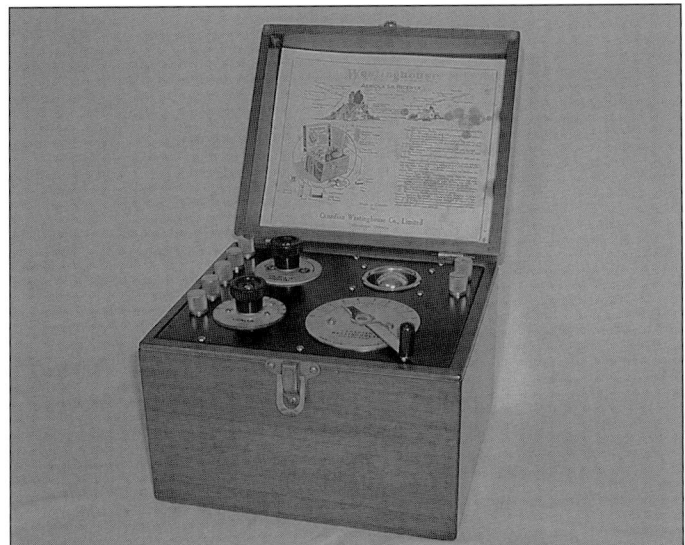

Fig. 2. Canadian Westinghouse Aereola Sr. (Robert MacIntyre collection)

Fig. 4. Canadian Westinghouse Radiola III (Robert MacIntyre collection)

hand edge, in contrast to the CGE which uses a sub-assembly of wafer capacitors. The contacts to the rotating coils in the Canadian Westinghouse and RCA models are through small coil springs apparently of bronze, while the contacts to the rotating coils in the CGE model are through short lengths of Litz wire fed through the hollow rotating shaft.

All indications here point to the likelihood that Canadian Westinghouse adopted the RCA version of the Radiola III except for the panel engraving. The CGE version contained some modifications. Does this mean that Westinghouse took the lead in manufacturing the Radiola III for RCA in the U.S.? Possibly so. Also, Canadian Westinghouse was the only subsidiary of the RCA family to sell a Balanced Amplifier in Canada, as far as we know. The RCC service manuals are not helpful in solving this puzzle. They contain a Canadian Westinghouse schematic which is identical to the RCA schematic shown in Riders. On

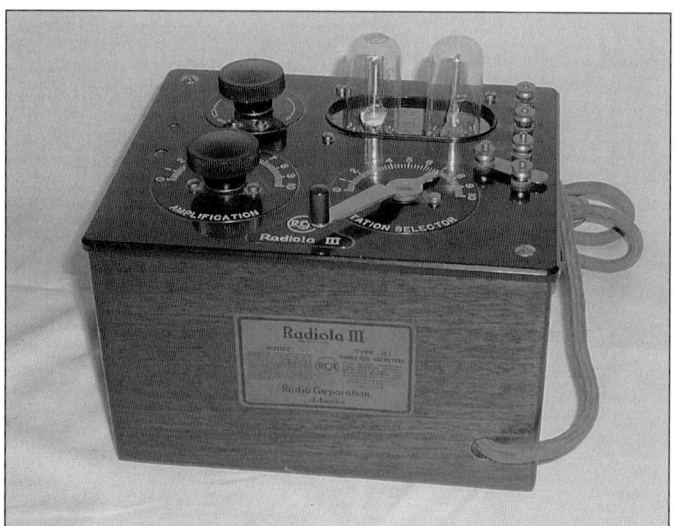

Fig. 5. RCA Radiola III (Robert MacIntyre collection)

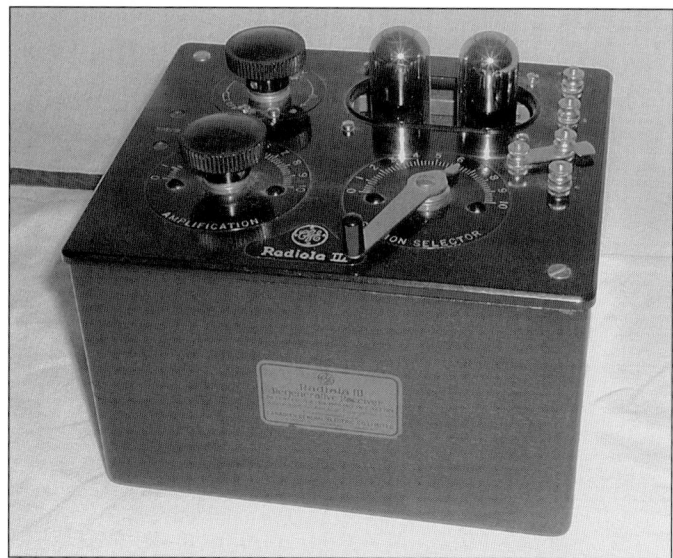

Fig. 6. CGE Radiola III (Robert MacIntyre collection)

Fig. 7. Components side of Canadian Westinghouse Radiola III (Robert MacIntyre collection)

Fig. 8. Components side of RCA Radiola III (Robert MacIntyre collection)

Fig. 9. Components side of CGE Radiola III (Robert MacIntyre collection)

the subject of the CGE circuit, they merely note that it corresponds to the Canadian Westinghouse circuit. Did Westinghouse manufacture the majority of, or all of the Radiola III's obtained by RCA? On balance it would appear so, and Barnum (1991, p. 144) specifically says that they were designed by Westinghouse.

The Radiola IIIA was also offered in Canada by both Westinghouse and GE subsidiaries. The RCA model is shown in Douglas (1991, p. 34) and in Bunis & Bunis (1992, p. 157). The CGE Radiola IIIA was finished in gold wash as was their Radiola III (see Fig. 10). The receiver used '199 tubes and inside used the same style of coil contacts, audio transformer and capacitors that distinguished the CGE Radiola III. The Canadian Westinghouse Radiola IIIA used WD-11 tubes, had nickel plated hardware, and in other respects looked like a Westinghouse product (see Fig. 11). A comparison of these two sets serves to reinforce the notion that Westinghouse was primarily responsible for the production of Radiola IIIA receivers for RCA, although GE may also have been involved. Were there some RCA Radiola III or IIIA models that used '199 tubes? None that we are aware of.

The Radiola Superheterodyne, AR812, was also sold by CGE and Canadian Westinghouse in Canada. My example indicates RCA on the escutcheon and on both dial plates (see Fig. 12). The RCA model is shown in Douglas (1991, p. 28) and Bunis & Bunis (1995, p. 200). It is only on the battery instructions behind the battery doors that the company CGE is identified (Fig. 13). There was also a version of this CGE set that had the CGE logo on the tuning dials, and the version shown in the Canadian Instruction Book shows a set identified this way. The RCC manuals do not contain a schematic for this receiver. The other information we have about its provenance is from John Anderson, who reports that its manual was printed in Schenectady and that there are 12 photo cards for this receiver in the Schenectady museum. The Canadian Westinghouse ad on page 142 also shows a Radiola Superheterodyne.

The Canadian Westinghouse model 8 (Fig. 14) is another superhet modeled loosely after the RCA Radiola 28 (which was also sold in Canada). The RCA Radiola 28 is shown in Douglas (1991, p. 41) and Bunis & Bunis (1992, p. 158). The model 8 schematic does not correspond to the RCA 28 schematic, and the model 8 cabinet is of solid hardwood, probably walnut, whereas the Radiola 28 cabinet is veneered. The Canadian Westinghouse model 8 does not correspond to any RCA model of which we are aware. In Table 2, the model 8 is compared to the RCA Radiola Super VIII. All of the other indicators of provenance of the Super VIII favor manufacture by GE.

The Canadian Westinghouse Radiola X appears to be identical to the RCA Radiola X shown in Douglas (1991, p. 30) and Bunis & Bunis (1995, p. 199), except for the medallion in the center of the front panel bearing the manufacturer's name (Fig. 15).

RCA Radiola 16 appears from Table 2 to have been manufactured by GE. The RCA Radiola 16 appears in Douglas (1991, p. 44) and in Bunis & Bunis (1995, p.

Fig. 10. CGE Radiola IIIA (Author's collection)

Fig. 11. Canadian Westinghouse Radiola IIIA (Canada Science and Technology Museum collection)

Fig. 12. CGE Radiola Super, model AR812 (Author's collection)

199). There is a corresponding Radiola 16 sold by CGE (see Fig. 16). On the bottom of the cabinet the CGE sticker appears to cover an RCA sticker of about the same size (not shown).

Canadian Westinghouse also made two receivers of types similar to the 16. One was called the Canadian Westinghouse 16 and was housed in a slightly larger cabinet than the CGE equivalent (Fig. 17). In other respects the CGE Radiola 16 and the Canadian Westinghouse 16 were very similar.

The other Canadian radio similar to the 16 was the Westinghouse Battery Operated. The Battery Operated appears equivalent to the previous

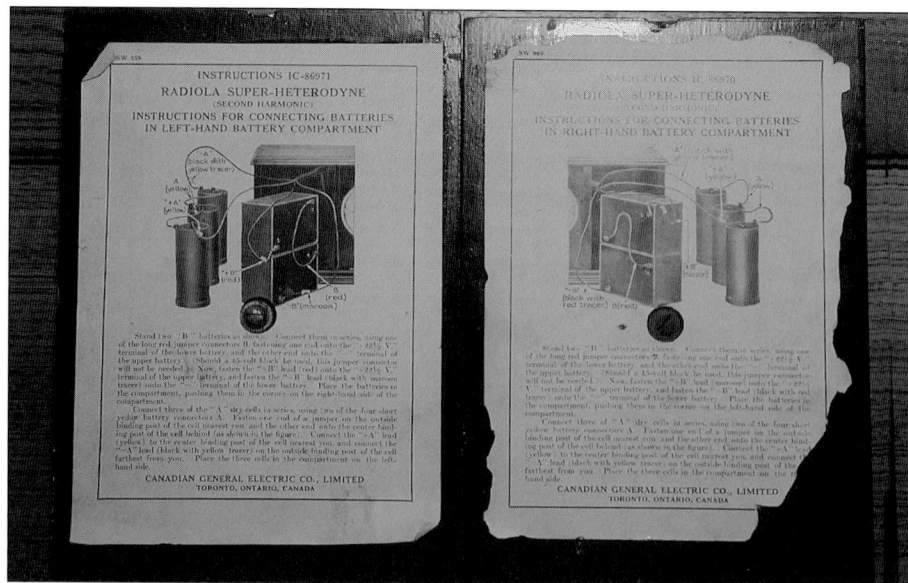

Fig. 13. CGE Radiola Super, model AR812, back side of battery doors. CGE is identified as the manufacturer on these labels. (Author's collection)

Fig. 15. Canadian Westinghouse Radiola X (Author's collection)

Fig. 14. Canadian Westinghouse model 8 (Canada Science and Technology Museum collection)

Fig. 16. CGE Radiola 16 (Robert MacIntyre collection)

two receivers but is contained in a box the same size as the Canadian Westinghouse 16 (Fig. 18). The likely origin of the name requires some explanation. Notice that the RCA Radiolas 16, 17 and 18 were all introduced around 1927. Radiolas 17 and 18 were light socket operated AC receivers. In Canada the pioneer in light socket operated radios was Ted Rogers (Chaplin, see chapter 4.1 above), who used the trade name "Rogers Batteryless". Apparently the name was not registered as a trademark because Canadian Westinghouse launched a receiver similar to the RCA Radiola 17 but called the Westinghouse Batteryless Receiver (Fig. 22 below). In contrast then, Canadian Westinghouse appears to have labeled the latter part of its production run of model 16 receivers as the Canadian Westinghouse Battery Operated.

To return to the likely manufacture of RCA receivers, there were CGE models of Radiolas 17 and 18. RCA Radiolas 17 and 18 are shown in Douglas (1991, p. 44) and Bunis & Bunis (1995, p. 199, Radiola 17 only). The CGE Radiola 18 pictured here (Fig. 19) is marked Canadian General Electric on its escutcheon (Fig. 20), but the paper tag under the cabinet covers what seems to be a dark brown RCA tag. Looking carefully at Fig. 21, the top edge of the dark sticker can be seen along the top of the CGE sticker. The indicators of manufacture of the RCA Radiola 17 and 18 in Table 2 seem to favor manufac-

Fig. 19. CGE Radiola 18 (Robert MacIntyre collection)

Fig. 17. Canadian Westinghouse model 16 (Canada Science and Technology Museum collection)

Fig. 20. Escutcheon of CGE Radiola 18 (Robert MacIntyre collection)

Fig. 18. Canadian Westinghouse Battery Operated (Canada Science and Technology Museum collection)

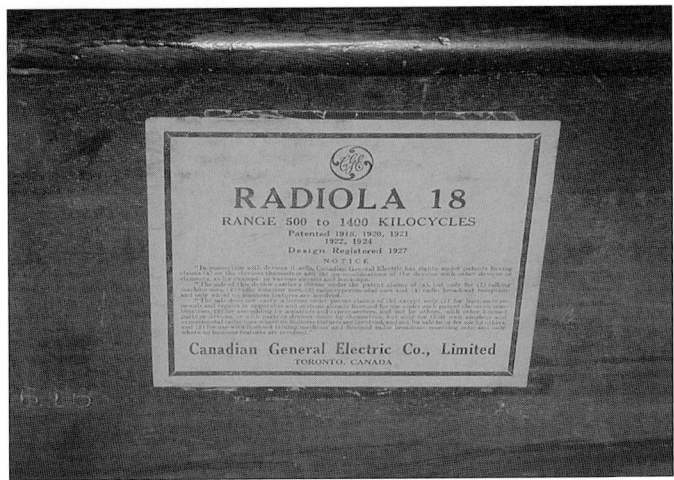

Fig. 21. Paper tags under the CGE Radiola 18 (Robert MacIntyre collection)

ture by General Electric. The fact that Canadian Westinghouse manufactured the Batteryless around the same time (Fig. 22), in a larger cabinet with a

Fig. 22. Canadian Westinghouse Batteryless Receiver (Canada Science and Technology Museum collection)

Fig. 23. Canadian Westinghouse Radiola Regenoflex (Canada Science and Technology Museum collection)

Fig. 24. CGE Radiola 51 (Author's collection)

somewhat different design that has no corresponding RCA model, encourages but does not prove the inference that no Radiolas 17 or 18 were manufactured for RCA by Westinghouse.

The RCA Radiola Regenoflex has no indicators in Table 2 about its manufacturing origin, and specifically none of the indicators that favor manufacture by GE (see Douglas, 1991, p. 32) and Bunis & Bunis (1992, p. 158). There was a Canadian Westinghouse Radiola Regenflex (Fig. 23) offered for sale in Canada which appears identical to the RCA model except for the engraved escutcheon which is part of the bakelite front panel. The existence of this model in Canada implies manufacture by Westinghouse for RCA.

RCA Radiola 20 (Douglas, 1991 pp. 40 and 43, and Bunis & Bunis, 1995, p. 199) appears from the indicators in Table 2 to have certainly been manufactured by GE. There was also a Canadian Westinghouse Radiola 20 sold in Canada, raising the possibility that Westinghouse also supplied some Radiola 20s to RCA.

Radiola 50 and 51 were apparently made by GE for RCA in the U.S. according to the indications cited in Table 2. For illustrations see Douglas (1991, pp. 44 and 47). Models of these receivers were also sold in Canada under the CGE label but not under the Canadian Westinghouse label (see CGE Radiola 51 in Fig. 24). The schematics of the RCA models in Riders and the CGE models in the RCC manuals correspond exactly. Finally, the indications for Radiolas 17 and 18, the table model equivalents of these receivers, are entirely parallel. They also were sold only by CGE in Canada.

The paper label for the CGE Radiola 51 is shown in Fig. 25. In fact there are two labels pasted one over the other on the shelf behind the receiver. The top label indicates that the receiver was made in Canada by CGE. The bottom label is shown where the corner of the upper one has been peeled back. The lower label indicates that the set was offered for sale by RCA.

These duplicate paper labels, found on Radiola 16, 18 and 51 in our possession and possibly on other models, indicate that likely complete models were exported to Canada, with Canadian escutcheon plates added either at the source or the destination. It seems unlikely that the addition of a name plate would allow these receivers to be identified as made in Canada under government Customs rules, but possibly if sub-assemblies were shipped, this would have been allowed. The models we encountered that

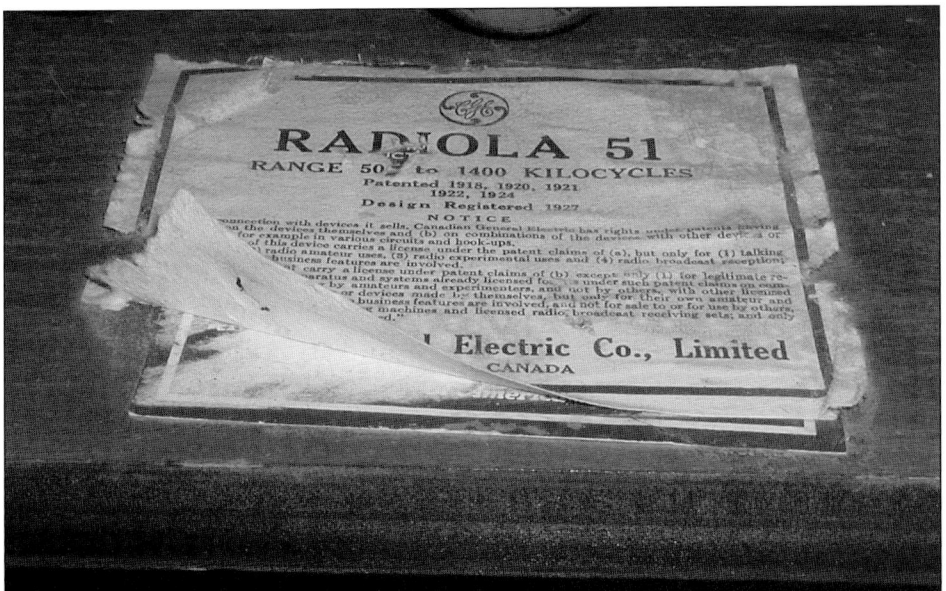

Fig. 25. Paper labels for the CGE Radiola 51 (Author's collection)

were so identified were all imported by CGE.

The Canadian Westinghouse 53 is a model advertised as having similar performance to the Radiola IIIA, but achieved with the use of only three tubes (see Fig. 26). The tubes were two UX-199s and one UX-120. No such corresponding set was sold by Westinghouse or RCA, and it is mentioned here only for completeness.

Canadian Westinghouse also offered at least three additional models similar to the Radiola 20. These were labeled the Canadian Westinghouse 55 (Fig. 27), the Canadian Westinghouse 57 Special

Fig. 26. Canadian Westinghouse 53 (Author's collection)

Fig. 28. Canadian Westinghouse 57 Special (Canada Science and Technology Museum collection)

Fig. 27. Canadian Westinghouse 55 (Robert MacIntyre collection)

Fig. 29. Canadian Westinghouse 60 (Canada Science and Technology Museum collection)

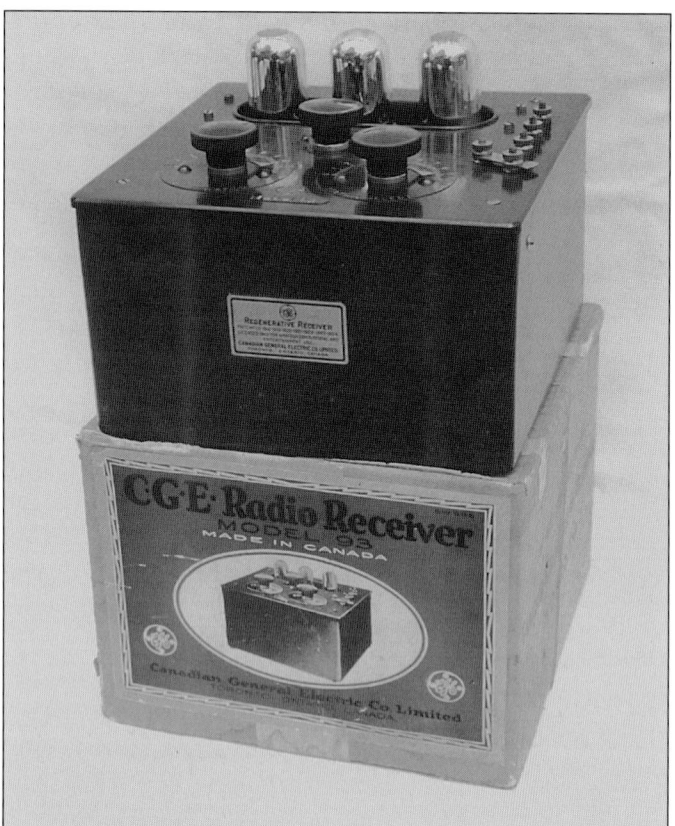

Fig. 30. CGE 93 (Canada Science and Technology Museum collection)

(Fig. 28), and the Canadian Westinghouse 60 (Fig. 29). Two of these were housed in cabinets similar to the RCA Radiola 20 cabinet, and the third was in a similar but less elaborate cabinet. It is not entirely clear why the designers at Canadian Westinghouse added this string of extra models, but their inclination to embellish on the designs handed to them by the parent company has been recognized.

The Canadian Westinghouse 55 does not correspond to the RCA Radiola 55. The respective schematics do not match. There was no receiver marked Radiola 55 sold in Canada.

There was a final set in this series, the CGE 93 (Fig. 30). This set did not correspond to an RCA model but rather to the Canadian Westinghouse 53, also apparently designed in Canada. As in the case of the CGE 53, this model is included only for completeness. It does not contribute to our indications of RCA manufacturing. It does, however, indicate that Canadian Westinghouse and CGE engineers were collaborating to some degree. The RCC manuals indicate that the circuits for the 53 and 93 correspond.

In the following chapter (9.2) model 205 is also mentioned as having been manufactured solely by CGE. Please refer to that chapter for a photo of model 205.

Summary

The essence of the information gleaned from the various sources is contained in Table 2 in the column headed "Made by:". There is essential agreement among the indicators on the subject of who made which receiver. The majority of them (38 in all) were made by GE and many fewer (17) were made by Westinghouse, with some overlap. Had we uncovered more detailed information from Westinghouse, the table might have been tilted at least some more in their favor. However, the information on Canadian models was not biased in this manner. The agreement that formed RCA suggested that 60% of manufacturing would be contributed by GE and 40% by Westinghouse. Counting just the number of models, the distribution is 69% from GE and 31% from Westinghouse.

Douglas (1991, p. 54) presented a table of sales volumes from RCA in the interval from 1925 to 1929. The early years are omitted. It would be possible to combine our results with theirs to see of the 60-40 allocation between GE and Westinghouse continued to be observed. In fact most of the Westinghouse models in Table 2 occur before 1925. This suggests that either Westinghouse was excluded from manufacturing for RCA as the years went by, or else our crediting of manufacturing to Westinghouse is seriously understated.

Conclusions

We identified from several sources, indications where in the 1920's Radiolas were manufactured for RCA by one but not both of its broadcast receiver subsidiaries. We also identified instances where receiver models were evidently manufactured by both of the subsidiaries. Overall though, receivers with single source origins appear to greatly outnumber receivers with dual sources, the stated goal of RCA after its initial two or three years of operation. Engineering difficulties and residual competitive attitudes of the two companies would likely have been the reasons preventing the full implementation of this goal.

We were limited in this line of inquiry of course by the eighty years of elapsed time, and also by the fact that we did not find equivalent information from both of the RCA member companies concerned. In the case of GE we found relatively detailed archives of receiver manufacturing in the Schencetady Museum, and a dedicated archivist in the person of Dr. John Anderson, himself formerly of General Electric. In the case of Westinghouse we found much less information. The prominent Westinghouse archivists in the AWA, Lauren Peckham and Bruce

Roloson, found that their records emphasized vacuum tube developments almost exclusively, and there does not appear to be an archive of Westinghouse manufacturing that is held today by a public museum or library. Unfortunately, too often corporations have a tendency to discard records that are more than a few years old.

Direct reference to broadcast receiver manufacturing by GE and Westinghouse is found for the first few years of RCA in Archer (1939). Beyond the first few years, this topic has never seemed to catch the attention of historians. It must be regarded as something of a mystery and at the same time unfortunate that the small and struggling manufacturing companies appeal more strongly to the imagination of radio historians than do the relative giants. As a consequence we are left with gaps in the historical record about which we know relatively little. The above account is an attempt to rectify one small part of that missing history.

REFERENCES

1. Anderson, J.M. Supplying tubes, sets, and people to RCA: the General Electric Connection. *The AWA Review*, volume 5, pp. 1-20, 1990.
2. Anderson, J.M. Personal communication, 2003.
3. Aitken, H.G.J. *The Continuous Wave, Technology and American Radio*, 1900-1932. Princeton, NJ: Princeton University Press, 1985.
4. Archer, G.L. *Big business and Radio*. New York: The American Historical Company, 1939.
5. Barnum, F.O., III. *"His Master's Voice" in America*. Camden, NJ: General Electric Company, 1991.
6. Bunis, M. and Bunis, S. *The Collector's Guide to Antique Radios*. Paduka, KY: Collector Books, 1991.
7. Bunis, M. and Bunis, S. *The Collector's Guide to Antique Radios*. Second Edition. Paduka, KY: Collector Books, 1992.
8. Bunis, M. and Bunis, S. *The Collector's Guide to Antique Radios*. Third Edition. Paduka, KY: Collector Books, 1995.
9. Chaplin, M. "Just plug in — then tune in" — the first commercial light-socket operated radio receivers with AC tubes from Rogers Ltd., Toronto, Canada. *The AWA Review*, volume 15, pages 147-178, 2002. Reprinted in this book as Chapter 4.
10. Douglas, S.J. *Inventing American Broadcasting 1899-1922*. Baltimore, MD: Johns Hopkins University Press, 1987.
11. Douglas, A. *Radio Manufacturers of the 1920's. Volume 3*. Vestal, NY: The Vestal Press, 1991.
12. Gutterman, L. *The Wisdom of Sarnoff and the World of RCA*. Beverly Hills, CA: The Wisdom Society, 1968.
13. Herron, E.A. *Miracle of the Air Waves: A History of Radio*. New York, NY: Julian Messner, 1969.
14. Howeth, L. *History of Communication-Electronics in the United States Navy*. Washington, DC: U.S. Government Printing Office, 1963.
15. Lewis, T.S.W. *Empire of the Air: The Men Who Made Radio*. New York: Edward Burlingame Books, 1991a.
16. Lewis, T. Life of David Sarnoff. Presented at the 30th Annual Conference of the Antique Wireless Association, Rochester, NY, Sept. 1991b.
17. Radio Corporation of America, *Radio Enters the Home*. New York, NY: Radio Corporation of America, 1922.

Major C.L. Richardson, Canadian Radio Engineer
by A.D. Challoner

Charles Lepel Richardson played a significant role in the early growth of the commercial and military electronics industry in this country. He was a practical engineer, as opposed to a brilliant inventor, whose technical standard, leadership and drive helped to advance our native industry. Above all, he was widely regarded as a fine gentleman.

Many of us will have in our collections one of the unique Canadian Radiola models produced in the mid-20s by Canadian General Electric, such as the Radiola 93, 193, or 205 or, perhaps the Radiola III or IIIA which used UV-199 tubes instead of WD-11s. How many, though, knew that C.L. Richardson as C.G.E.'s first radio engineer was responsible for their development? How many of our Canadian members who are also members of the Institute of Electrical and Electronics Engineers of today knew that C.L. Richardson was the founding chairman of the Canadian Section of the original Institute of Radio Engineers which later grew into the IEEE?

Fig. 1. Major Charles Lepel Richardson, 1891-1978.

Professor C.G. Richardson

Charles Gordon Richardson immigrated to Canada with his parents in 1876, and was to lead a very colourful and successful career as a Professor of Chemistry and later as an entrepreneur in the metal ore exploration and mining business in both Canada and the U.S.A.

On completion of a special course in Chemistry and Geology at the School of Practical Science, University of Toronto in 1880 he joined Dr. Andrew Smith's Ontario Veterinary College in Toronto as Professor of Chemistry. In addition to lecturing at the college, Professor Richardson gave many public lectures and demonstrations — a tradition which was to be carried on by his son. One series dealt with "water, heat, formation of snow and gases". Some were travelogues richly illustrated with novel stereoscopic slide projections, e.g., "A Trip to India".

In 1886, at the age of 26 he married Helena Rattray, a young Scottish immigrant, and their first and only child, Charles Lepel was born in Toronto on 7 June 1891.

Charles Gordon, in 1898, seized an opportunity to establish the New York Mantle Co. selling gas mantles under contract to the city of New York. He then reorganized this company in 1900 to form the New York Chemical Refining Co. specializing in refining zinc ores at laboratories in Waverly N.J. About 1904 he moved the head office to Elizabeth N.J., where Charlie's public school education was continued. It was during this period that he also learned several practical lessons from his father.

It seems that Charlie was permitted on occasion to drive his father's Stanley Steamer but was cautioned about always keeping the water tank filled. Sometimes when he was in a hurry and had not checked the water level, the Steamer would stall. When the water was replenished, though, it would restart. Now one afternoon his father was at work without the automobile; so he decided to drive to Hackensack, N.J. Once again it stalled, but, this time he could not restart it by simply adding water. After several futile attempts he finally asked a nearby farmer if he would tow him with his horses. And so it happened that Charlie arrived home for dinner, late and rather unceremoniously in the Stanley

Steamer which wasn't any longer a horseless carriage. Well, to be sure, his father took full advantage of the situation and lectured him on the proper way to maintain equipment, in particular, and on life in general! He reckoned that Charlie must learn the importance of maintenance by first crawling under the Steamer and removing its boiler. When he had finally removed it after many hours of difficult labour his father instructed him to remove each and every piece of tubing and ream out the hardened deposits which had collected as the result of his continual negligence. Well, when that episode was finally over he had indeed learned a very practical lesson which he later claimed completely altered his attitude toward machinery.

Wireless Days

Somewhat of a stir was created in Elizabeth, N.J., the day young Charles erected his first aerial. The police investigated and reporters interviewed and the following headline appeared in the Tuesday evening edition of the *Elizabeth Daily Journal* (5 Oct., 1909):

"ELIZABETH BOY RECEIVES
WIRELESS NEWS AT HOME"

His home-built wireless system was reported to be able to receive messages from a distance of 300 miles and send out communications ten or fifteen. Most people of the day thought Marconi was the only man who could receive messages out of the air. His station comprised a coherer with potentiometer supply, a 100 ohm telephone receiver, fixed and variable condensers, two dry batteries, telegraph key, induction coil, spark gap and helix coil.

His education in such matters appears to have been derived from studies in electrical and mechanical engineering (the only formal record is an honors diploma in Science from Pingry College preparatory school dated June 1913), magazines on electricity and wireless telegraphy and fellow amateur experimenters in the American Wireless Association. His father no doubt encouraged his interest in such matters.

It then happened that Professor Richardson returned with his family to Toronto to organize the Dominion Metals Co. Ltd. at 137 Jarvis St. to establish a pilot plant for development of a process for the reduction of Nipissing ores containing cobalt, nickel and copper mixed with large amounts of iron and arsenic. The process was soon proven successful and larger scale production was undertaken in the old Dominion Radiator Plant on Dufferin St. Young Charles helped out in the plant as Assistant Engineer and also undertook training as a Sapper in the University Section of the 2nd Coy., Canadian Engineers. The latter activity was to lead to an interesting episode of military wireless.

During the early part of 1910 on his own initiative, C.L. Richardson developed a wireless pack set, weighing about 92 lb. with a sectional mast of 32 ft. length. The apparatus fitted a suitcase and contained a complete receiving and transmitting set capable of operating over a distance of some two or three miles, which was considerable in those days. He was short-

Fig. 2. Station IXCR on the day it was opened, Nov. 25, 1913. On the right is the transmitting set, showing the oil-immersed condensers, 3-spark coil, operated from power mains through an electrolytic interrupter. The rotary gap is on top of the coil, and behind it is the tuning helix. Above is a giesler tube indicator. On the desk to the left is the receiving equipment, consisting of tuning coil, condenser, crystal detector, fixed condenser and change-over switch. On the wall at the rear is the power control board.

Fig. 3. Interior views of Station IXCR at Welland, Ont. These show the improvements made in the installations. On the right is the transmitter, surmounted by the change-over switch, while at the left is the receiving cabinet operating on crystal. On top of the cabinet is a copy of a Marconi detector.

ly promoted to Sergeant and assigned a small detachment for wireless instruction and by the end of that year, two more improved pack sets were constructed. They were featured during a sham fight at Lambton Mills in which a large Packard truck was used with two short masts erected at both ends with aerials strung between on spreaders. This was the first time that wireless was ever used by Canadian Militia in manoeuvres. Col. Maunsell from H.Q. at Ottawa made an inspection of the unit and was no doubt impressed when bombs from the moving truck were exploded in Toronto. Word soon spread and organizers of the Annual Motor Show in Toronto requested its presence at the March 1911 show. Newspaper accounts of the day showed tremendous interest in the Packard wireless truck. So successful was the exhibit that it was repeated at the Montreal Motor Show later that month. And it is not surprising that on 28 Apr. 1911, C.L. Richardson received his first contract from the Dept. of Militia and Defence for the construction of two identical pack sets for use at Petawawa Camp, Ontario.

Charles continued to give lectures in wireless at the Armouries, the Military Institute and the Dominion School of Telegraphy, Queen St. E., Toronto until the beginning of 1913 when his father moved the business to Welland Ontario. There under the name of Metals Chemical Co., business was booming. Starting with a processing capability of 15 tons of ore per day it was soon stepped up to 50 tons to supply cobalt and nickel to the U.K., U.S. and Japan during WWI. Young Charles was no doubt quite involved as Chief Engineer for the operation and had to drop wireless for a time.

Sure enough, though, the old urge started and a new station, IXCR, was opened on November 25th 1913, notice of which appeared in the Welland and Toronto papers. This rig already exhibited his characteristic care and craftsmanship, as can be seen in the accompanying photographs (Figures 2 and 3).

In November 1916 his health broke down from lead and arsenic poisoning forcing a four-month period of recuperation in Bermuda. Then due to the possible entry of the U.S. in the war he was sent to help his father establish a similar plant, Niasco Chemical Co., at New Market, N.J. in 1917; this later became the Cobalt Chemical Co. in 1920.

Owing to another break in health he was forced to give up chemical work entirely in 1921, and then fell back on wireless for employment. He was engaged as Chief Radio Instructor at the American Airways Training Schools Inc., College Point N.Y., a federal aviation school for war veterans.

Also during these post-war days he pursued a sideline interest in the theatre. He secured a small part as an officer returning from the war in Norma Talmadge's moving picture *"Smilin' Through"*; but, as parts became scarce he became a projectionist just to keep his hand in. He ended up at Proctor's Jersey Street Theater and when they changed from vaudeville to stock he remained and merged into a character actor and doubled as a lighting expert for the Proctor Players Company. His flair for the dramatic no doubt accounted for his later popularity as a public speaker.

The Radio Boom

On the advent of radio broadcasting, C.L. Richardson joined forces with an old school chum, Marcus Whitehead for the manufacture of radio apparatus at Elizabeth, N.J., under the name

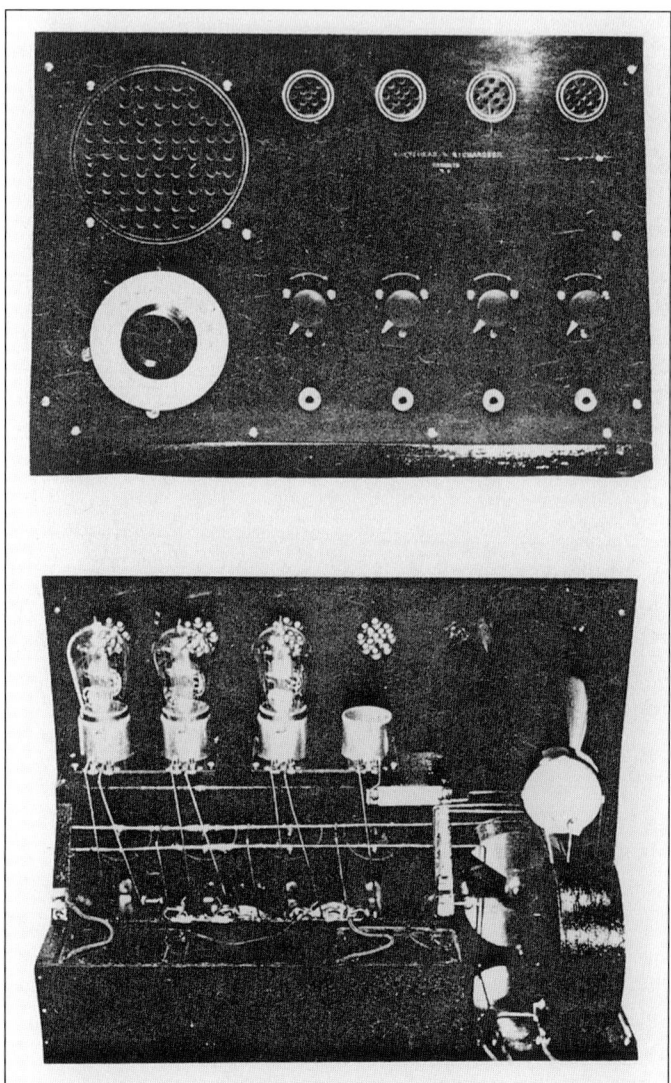

Fig. 4. Audiola A-300, as manufactured by Whitehead & Richardson for John Wanamaker Co., Marshall Field Co., Lord & Taylor and Gimbel Brothers Co.

Whitehead and Richardson. From about February 1922 to the end of 1923 a prodigious quantity of custom and short production run radios were made. They developed a reputation for high quality products, supplying John Wanamaker of Philadelphia, Marshal Field of Chicago and many other firms. Also some notable installations were made for the Union Club of New York, Mr. Reginald Vanderbilt and Maxim's Restaurant to name a few (Figure 4).

Perhaps sensing a greater freedom and opportunity, perhaps just longing for his 'home' town he decided to pack up his operation in Elizabeth and return to Toronto in January of 1924. The radio boom which was well established in the U.S. was just starting to mount in Canada, so opportunity abounded. He set himself up as a Radio Engineer in an office and shop on Clinton St. and because of his relatively extensive wireless and radio experience obtained in the more seasoned U.S. climate he was in very high demand. His Radio Service Bureau offered a rather novel monthly inspection service as well as regular repair work. He designed and built prototype receivers for Simpsons and Eatons Department Stores. He was sought after by various magazines of the day, such as, *Radio News of Canada* and *Radio Sales*, to write semi-technical articles; the latter appointed him Technical Editor. Of course with each introduction in the print medium there would be an impressive curriculum vitae. And so it was no small wonder that in September 1924 C.G.E., which was just launching its radio business, offered C.L. Richardson the key position of Radio Engineer reporting to Oscar E. Forrest, Manager of Engineering at the Ward Street Works. He was instrumental in setting up the Radio Engineering Department with its attendant laboratory and was responsible for the design, manufacturing specifications and drawings of all Radiola receivers produced by C.G.E. He was also a member of the Design Committee at General Electric Co., Schenectady, N.Y. While C.G.E. made selected sets of American RCA design, the Canadian company did design and develop several sets more suitable to the increased distances of the Canadian field and to the locally available parts and materials.

His department comprised three assistant radio engineers and three draftsmen; the radio manufacturing division employed some 470 workers during the peak months from June to December. Radio operations continued at C.G.E. until 1928 when RCA Victor of Montreal was set up to manufacture for Westinghouse, C.G.E. and themselves, along the lines of the American RCA. By Christmas of 1924 C.G.E.

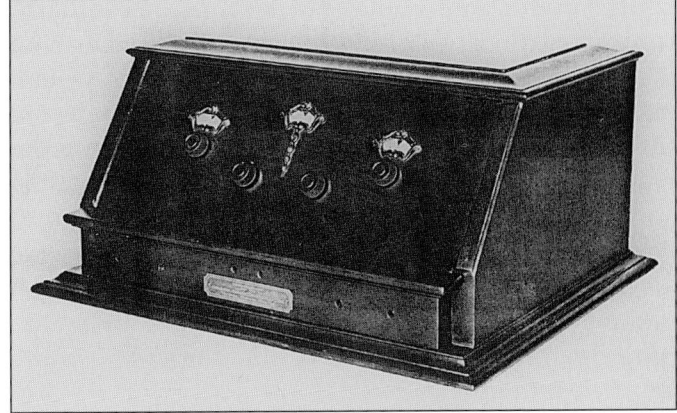

Fig. 5. C.G.E. Model 205, 5-tube TRF

were producing Radiola III's and IIIA's from scratch, using their own dies, moulds and tools. In the 1925-26 period they produced the following products on a similar home-made basis: Radiola Super VIII, Radiola Superheterodyne, Radiolas 24, 25, 28, 93, 205 and loudspeakers 100, 102, 104 and UZ1325. The Radiola 205 TRF receiver was particularly pleasing to C.L. Richardson as it was 100% C.G.E. designed and manufactured. It comprised one TRF stage followed by detector and three stages of audio amplification (Figure 5).

Of the many problems posed to the Radio Engineering Department, the following has become known as somewhat of a classic:

"Coils for a radio frequency amplifier were wound on a special winder and immediately dipped in a mixture containing ingredients having a certain dielectric constant. The purpose of this dip was to stiffen the coil for handling and to impregnate against moisture. After assembly into the amplifier unit a dielectric compensator was attached to represent a value when the whole coil system was immersed and solidified, after all adjustments had been performed. The compensator was on only during the period of adjustment and test, after which it was removed and the coil system immersed and sealed.

One day it was discovered that all amplifiers were being rejected after immersion and an immediate investigation was instituted. The shop processes were checked and then the test methods. Everything appeared O.K. in every way right up to the immersion of the coil system. A complete analysis of the immersion mixture as well as material entering manufacturing disclosed nothing of a troublesome nature. The output of the department was at a standstill and the days lengthened into weeks and finally, three went by and no answer. The engineering staff worked day and night and constructed new amplifiers from the ground up which passed O.K., but shop production could not get by. Eventually, through patient watching himself, C.L. Richardson discovered that one of the girl operators who wound and dipped coils had been using a pet mixture to dip the coils instead of the proper preparation from the main tank. She found that adding a dash of cleaning solvent reduced the tendency of the mixture to form lumps! She had dipped some 30,000 coils which were stocked two months previous and which the shop had just run into. The coils were salvaged by re-dipping operations, but think of the worry and three weeks lost time."

"Eternal Vigilance is the Price of Quality" — C.L. Richardson

On October 2nd 1925 an important step in the history of Canadian radio activity was taken when C.L. Richardson organized a Canadian Section of the IRE at a meeting held in the auditorium of the Ward Street Works. The list of those present was quite impressive. At the head table, along with C.L.R. were: Don McNicol, VP IRE, Cmdr. C.P. Edwards, Dept. of Marine and Fisheries, E.S. Mallet, C.G.E. Works Manager, Ward St. Div.; Profs. W.H. Price, T.R. Rosebrugh, U of T; D. Carlos, past president, Toronto Section, AIEE. Other notables among the 53 present were: W.F. Choat, Westinghouse Canada Ltd.; Chas. A. Lowry, H.M. Kipp Co; H. Lewis, Radio News of Canada, E. Swan, CKEY; A.M. Patience and A. Oxley, RCA Victor Co., Montreal; and G. Pipe, Rogers Majestic Co., Toronto.

In his address to the meeting, C.L. Richardson pointed out that there were at that time some 60 Canadian members of the IRE and an estimated 250 engineers actively engaged in radio design and construction in Canada as well as many who were connected with various government departments. He stated the object of the Canadian Section was to bring these men together in order that they may perform the same functions for the radio industry in Canada that had been performed in the United States by the American Institute, such as standardization of radio apparatus and the discussion of technical matters of interest to all. Also the Canadian Section it was said would provide in winter months popular illustrated lectures on technical matters presenting them in simple language for the benefit of the general public.

After the Radio Boom

When the University of Toronto heard that he was leaving C.G.E. in 1928, they offered Charles Richardson a professorial appointment, but he declined the offer in favour of a more lucrative and independent career as an electronics consultant. Until the great stock market crash in 1929 he remained in the radio field, associated with various firms such as Splitdorf of Toronto. He then turned to sound and film engineering and was associated with Dominion Film Labs, Film and Slide of Canada and later Screen & Sound Services Ltd. until about 1935 when he went completely independent. During the early '30s he set up automatic equipment for developing and printing 35 mm motion picture film, installed recording studios and worked on various aspects of sound equipment development and production. Between 1935 and 1940 he built many high quality custom sound installations for various companies and organizations, including: General Motors, Imperial Oil, Consumer's Gas and Conklin and Garrett Ltd. Fortunately, he was also an accomplished photographer and has kept pictorial records of many of his productions.

In 1940 he was assigned as training officer to the Canadian Corps of Signals at Vimy Barracks, Kingston, Ontario. He was then attached to the Directorate of Communications and Fire Control under Brig. H.E. Tabor. He was concerned with the technical requirements of all communications equipment to be manufactured in Canada, including radar, and liaison with the British and U.S. authorities. He spent an interesting stint in Kapuskasing, Ontario in January, 1942, in charge of the first cold weather tests of Canadian and U.K. communications equipment for use on the Russian front. Among the gear tested was the famous No. 19

Fig. 6. Wireless Set #19, Mk II. R.C.A. Victor & Northern Electric Co. Ltd. were manufacturers.

tank set, which, after the war, became the workhorse of the Canadian amateur (Figure 6).

His next appointment was to the Inspection Board of U.K. & Canada (I.B.U.K. & Can.), the agency for the inspection, test and acceptance of all electrical and electronics equipment required by the armed forces during WWII. He set up the Ontario section of the Signals Inspectorate, with some 183 personnel deployed in the Bonded Stores at each contractor's plant. It was his duty to resolve any problems between the Contractor and Bond, as well as to administer the organization and confer with Ottawa H.Q. on acceptance requirements. His high personal standards no doubt suited him well for this position and won him the immediate respect of various contractors. He was promoted to the rank of Major in 1945 and transferred to Ottawa to head the Signals Inspectorate, overseeing the Ontario and Quebec sections as well as the Ottawa H.Q. staff. The I.B.U.K & Can. completed its function in 1947.

For a few years following he was a consultant in private industry, generally associated with General Radionics Ltd; the first car radio sold by Canadian Tire Corp. was designed and developed under his guidance.

His seniority and rapport with the Canadian electronics industry, together with his experience with the I.B.U.K. & Can., led naturally to employment as Production Officer for the Electronics Branch of the Dept. of Defence Production in Toronto. It was his duty to inspect and assess production facilities of proposed government contractors, assist in solving technical and production problems arising during production and to confer with the Inspection Services on problems arising with contractors. As one comes to expect with the endeavors of C.L.R., when the closing of the Toronto office was announced in 1955 he received letter after letter from many firms with which he had dealt, expressing their regret and sincere appreciation of his kindly and valuable assistance with their problems.

In the years that followed he continued with some consulting work related to crystal manufacture and sound equipment, but gradually attended to more aesthetic matters. He was a lodge man and organist and took up the hobbies of HiFi, photography and genealogy.

Charles Richardson had an old Austin automobile, but, unlike a certain Stanley Steamer of years gone by, it was maintained to perfection. If he had occasion to drive the venerable Austin in the winter he would, on returning home an no matter what the time or temperature, thoroughly wash down the outside in his driveway; he would then roll it into the safety of his garage where a thorough scrubbing of the underside would complete the ritual. For such was the nature of Charles Lepel Richardson.

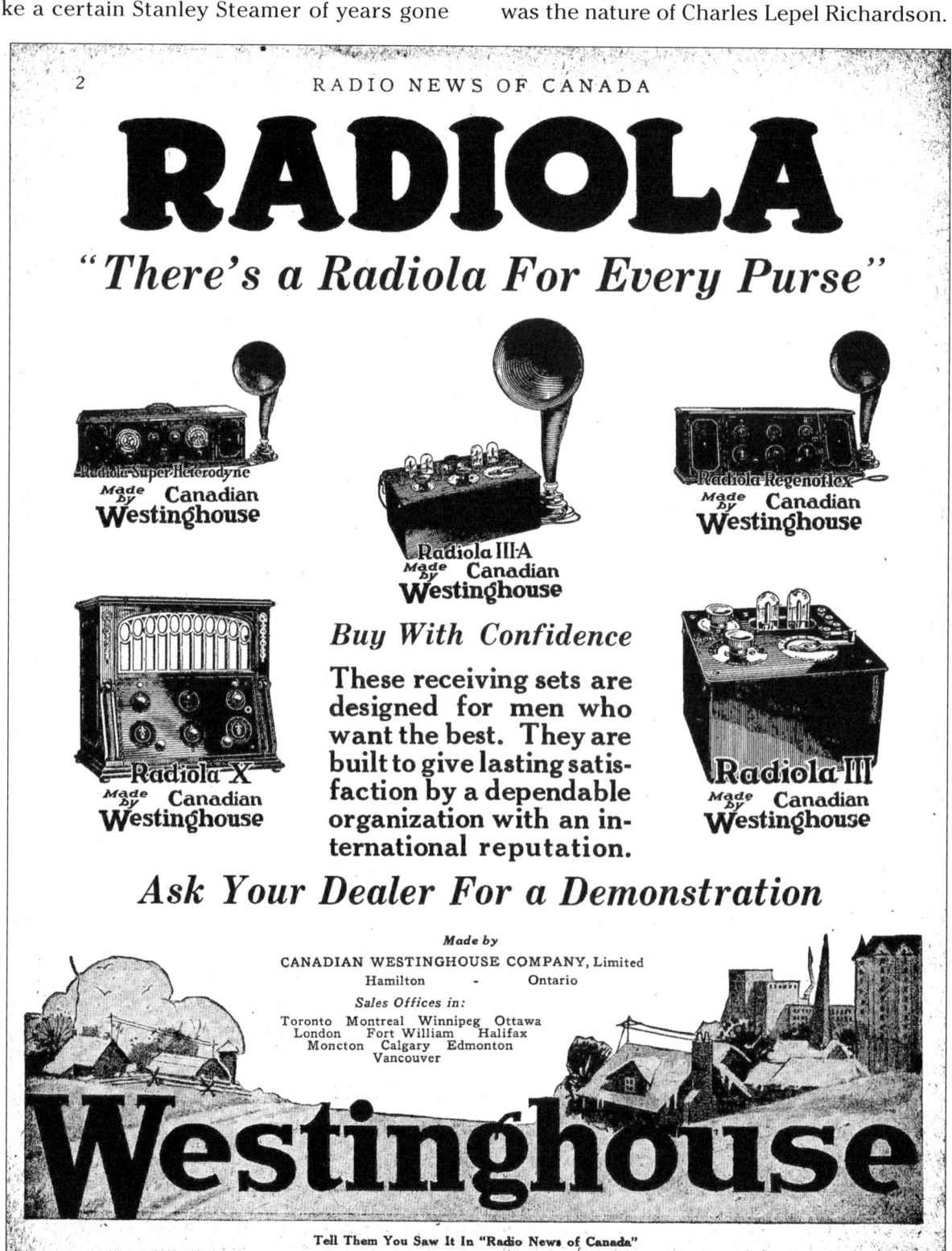

~10~
Starting a Radio Company in the 1940's

We close with the story of Thorcraft Radio, a small manufacturer of somewhat more recent times, whose history is not very different from those smaller early companies already described, and from the many smaller companies not mentioned here.

Hank Thorkelsson did not get into the radio manufacturing business full-time until 1947. In this he was clearly near the end of the vacuum tube radio manufacturing business. The evolution of his business was surrounded by some of the same problems that had plagued his predecessors. One was his apparent lack of access to patents. He had financial problems, in his case a fire from which his company never really recovered.

Thorkelsson designed a television, but his company did not have the financial strength to set up production for this, a logical next step for a radio manufacturing business in the 1950's.

Hank's real appeal to the radio collecting community was that when many of us were just getting started, he was still with us. He seemed interested in the radio collecting fraternity, and was ready to regale us with stories of the early days. He connected with collectors in both Winnipeg and Vancouver. He is fondly remembered.

Hank Thorkelsson and the Thorcraft Radio Co.
By Robert P. Murray

Haukur (Hank) Thorkelsson was born in Ashern, Manitoba in February of 1913. He started working for his father's trucking firm in the late 1920's. The company was called Ashern Freighters and is still going. Hank started driving trucks when he was 14 years old. When he was 20 he built his first receiver. It was a battery set because there was no hydro electric power in the area around Ashern in 1933, nor was there for some years afterward (Figure 1).

Hank started in the radio sales business in Ashern in 1936, first selling Rogers farm battery sets and later switching to RCA. While with RCA, he achieved the highest sales of battery sets in Manitoba. At that time, Hank discovered that in order to be successful in radio sales, he also had to learn radio repair. He never took a radio course as such, but learned from books and from practice. In 1940 he was hired by Eaton's in Winnipeg, first in their radio repair shop and later on outside repair calls. It was there that he met Gar Gillis who later set up in business with him.

Hank and Gar were the Eaton's outside repair crew for a couple of years. They each drove an Eaton's repair truck. They would report for work at the beginning of the day, and each get their day's assignment of calls. They soon discovered that the workload was easy to manage, and they would head off and meet for coffee out of sight of the store. They could linger at coffee during the morning, and still get all their calls done if they started at 1:00 pm and worked really hard.

Fig. 1. Hank Thorkelsson holding a Thorcraft Model 5B4T

Hank and Gar left Eaton's in 1942 and set up a service shop at 130 Osborne St. in downtown Winnipeg. After a couple of years Gar left to form his own electronics business, manufacturing amplifiers. Hank started manufacturing receivers on the side. The first ones didn't have his name on them. Then a few were labelled "Thorkelson Radio Company." An early one was the model D2 designed for 32 volt farm operation (Figure 3).

In 1947 Hank got out of the radio service business and went entirely into manufacturing. He moved into premises first at 592 Erin Street and later in St. Boniface. In 1950 the company suffered a major fire. They had just taken delivery of a shipment of 5,000 tubes. Hank remembers that the cartons burnt off the tubes, and they rolled out in the stream of water and into the sewer. The fire department arrived and flooded the building with water. When the wiring in the building shorted out, the fire flared up again. They lost $50,000 worth of inventory and only had $10,000 of insurance coverage. The company struggled on, but never recovered financially from the fire.

In the 1950's Thorcraft made radios under the Arcadia and Coronado labels for MacLeod's, under the Aurora and Viking labels for Eaton's, and under the Seranader label for Simpsons-Sears. They also made a battery set called the Northern Messenger for The Bay — it was sold in the remote North. These were all Thorcraft designs sold under the various labels. You won't find any of them in your RCC manuals though. Hank says that although he was approached, he never chose to send in any of the schematics.

The company did a test run of television sets in the early 1950's. They assembled 25 black and white sets. They would have been sold under the Thorcraft name, but it was found that the production was too expensive to set up. The company peaked in size around 1952-53 with 25 employees. In total they built over 10,000 radios. These included battery radios, floor models, 32 volt farm radios, school radios and hospital radios. They were sold nationally in Canada. Hank had a friend who became national manager of Mallory in Canada and was his unofficial representative. He sold a lot of radios in Newfoundland through this connection. His Newfoundland distributor was James McGrath who was in the furniture business, and also a member of parliament. The company didn't use much advertising.

In 1957 the company went bankrupt, but then started up again. Hank finally closed the business in 1964 and moved to Vancouver. There he worked briefly for a partnership that had bought out the

Fig. 2.   Thorcraft model B71

Fig. 4.   Thorcraft model 35ET

Fig. 3.   Thorcraft model D2

Fig. 5.   Thorcraft model 5A4T

Table 1
Partial list of the many Thorcraft radios manufactured.

Model	Power	Tube lineup	Notes
D2	32 vdc	6SA7,6SK7,6SQ7,6J5,25L6(2)	farm radio for windcharger
B62	32 vdc	6SA7,6SK7,6SQ7,6J5,25L6(2)	"The Voice of the West"
B66	32 vdc		6 tube reflex, push-pull output
B71	32 vdc	6SK7,6SA7,6SK7,6SQ7,12SN7,25L6(2)	
55BT	1.5 & 90v		5 tubes
35ET*	110 vac/dc	12BE6,12BA6,12AV6,50L6,35Z5	Hybrid with 7-pin and octal tubes/schematic
55ET	110 vac/dc	5 tube superhet	
55ETDL	110 vac/dc	5 tube superhet	two speakers
76EHC	110 vac/dc		Hospital radio with. Dahlberg pillow speaker
85ET	110 vac/dc	12BE6,12BA6,12AV6,50C5,35W4	
85EMTS	110 vac/dc	12BE6,12BA6,12AV6,50L6,35Z5	two speakers
4B9T	1.5 & 90v	4 tubes	
5A4T	110 vac/dc	same as 85ET	
5A4T2	110 vac/dc	same as 85ET	two speakers
5B4T	110 vac/dc	same as 85ET	phono jack
5B4T2	110 vac/dc	12BE6,12BA6,12AV6,35C5(2)	push-pull output, silicon rectifier, two speakers
5B4TE	110 vac/dc	same as 85ET	utility version grey cabinet
6A6T	110 vac/dc	6 tubes (extra rf stage)	5 x 7" speaker
6A7T	1.5 & 90v	6 tubes	5" speaker
7A8TS	110 vac/dc	12BE6,12BA6(2),12AV6(2), 35C5(2),selenium rectifier	school radio, grey cabinet, 6 x 9" speaker dated Jan. 1960
7B8T2	110 vac/dc	same as 7A8TS	two speakers, 5 x 7"
6B10K4	110 vac/dc	6 tubes, rectifier diode	stereo radio phonograph, floor model
8B12K6	110 vac/dc	8 tubes, rectifier	stereo phono, without radio
11B11K6	110 vac/dc		same as 8B12K6 but with radio

*Note: Model 35ET was also produced as the Keystone Royal, mfg. exclusively for Keystone Supply Co. Ltd., Winnipeg, Man. by Thorcraft Ltd., Winnipeg.

Tenco importing agency and were beginning to manufacture amplifiers. In six months he was on his own again, manufacturing amplifiers under the name of Grant, his son's name. He was still producing some Grant amplifiers at Grant Electronics Ltd.

In 1994 when this article was first written, he and his long time friend Gar Gillis were both still in the vacuum tube amplifier business. They collaborated again to some extent. Now and again they repaired an antique radio. They were both going concerns in their 80's. Hank has since passed away in Vancouver.

Thorcraft Broadcast Receivers

As mentioned above, these radios are not listed in RCC manuals. Printed information on the models made, their year of manufacture, power source, purpose, etc. is not available. Table 1 is a partial list, based on several sources.

All of the above models were provided with

Fig. 6. Thorcraft print ad

veneered wood cabinets, in walnut, mahogany or light oak finishes, except for models 5B4TE and 7A8TS. These were in wood cabinets, but sprayed with a grey spackle plastic finish. Selenium rectifiers were used for a short time but Hank found them unreliable. As soon as silicon rectifiers were available, they were adopted.

Thorcraft radios used a number of circuit variations to meet specific purposes. They were not bound by the "all-American five" ac/dc set that was the mainstay of so many manufacturers in this period. They showed some originality. It also makes them interesting radios to collect.

Acknowledgements

Thanks are due to Hank Thorkelsson and Gar Gillies whose recollections are the basis of this article. Thanks also to collector Ken Blair who first introduced me to Thorcraft Radio. An appeal for information on additional models brought responses from Tom Dewing, Herb Fredrickson, Stan Marlin, George Phelps and Roy Sawley. All of the models photographed in this article are now in the Canada Science and Technology Museum.

Fig. 7. Thorcraft print ad

Epilogue

As the years wore on, companies that were bigger and could absorb occasional losses continued to survive in one form or another. After surviving the Great Depression, Canadian Marconi, Northern Electric, Rogers Majestic, Canadian General Electric, Canadian Westinghouse and RCA Canada continued to dominate the market. Others tended to remain small and their place in the radio market remained fragile. Grimes/Electrohome, an intermediate sized company, succeeded. Thorcraft, a later start-up company without patent access, foundered.

Some of the effect of the concentration into RCA spilled over into Canada. RCA Canada was established here in 1928. Although I have not studied it closely, their mandate appears to have been to provide receivers that had been designed in the US market, to a Canadian audience. They did not work to undercut the business that was already in Canada and served by their subsidiaries. Canadian Westinghouse and CGE continued to offer radios in a market in which they were well-established, and RCA Canada co-existed with them.

An earlier, somewhat similar rearrangement of alliances was in the case of the Victor Talking Machine Co. Victor had a trial affiliation with the Northern Electric Company for the purpose of enhancing radio sales, we presume, about 1925. Northern provided the chassis and Victor contributed the cabinet. This alliance continued for about two years. When RCA in the US decided to merge with Victor to form RCA Victor, the word must have been passed here that the Victor/Northern Electric collaboration would be unpopular in the RCA Victor corporate office. I have examined Northern Electric documents carefully and not found any indication of this, but it appears that a corporate change in the US affected inter-company agreements in Canada.

The US and Canadian radio markets interacted in many ways, as we have seen. The Canadian and American Marconi companies diverged, with American Marconi (RCA) becoming much more dominant in its market than Canadian Marconi ever did. Part of this was by US government design, and part due to the personalities of its officers. The inter-corporate agreement that constrained Northern Electric was very different than that which constrained Western Electric with respect to radio business. In this case the Canadian company got a much broader mandate. The Rogers company discovered an early AC-operated tube under development at the US Westinghouse company laboratories. They bought the Canadian rights to it, made some improvements to its design, and aggressively brought it to market in Canada. Their acumen in those early days has translated into a vigorous corporate giant in domestic cable, video rentals, print media and cellular phone businesses today. The playing field at the Canada/US border has sometimes tilted one way and sometimes the other.

Many of the large companies in this history are still in business. I think that none of them still makes broadcast receivers. Their products are broadly diversified among military contracts, data networking, cable and print distribution of news and entertainment, etc. What broadcast receivers exist are made predominantly in Asia. There is no broadcast receiver manufacturing industry in Canada, nor has there been for many years.

In the 21st century the story is almost complete. There is talk of conversion to digital broadcasting. At some point in the not too distant future, AM radios will simply no longer work to bring you broadcasting. This change will be the result of technical progress which transcends national borders. Its intent is to bring higher quality reception. There is no hint here of a bias against the Canadian industry.

Finally, a word about collecting radios, the place where this particular history began. Collecting is a passion of many, and radio collecting is certainly popular. One problem with collecting radios is that there are simply too many of them. One needs to have a narrower focus. It is worth considering that you can arrange a focus where the focus itself is interesting to others, not only the collection. It turned out in my case that collecting Canadian radios of the first decade of broadcasting was the focus. There are a number of Canadian collectors with the same focus, which is a useful realization, and we help each other. This focus caught the attention of the Canada Science and Technology Museum- also a good thing, I think. Finally it led to this book. It gave me materials to illustrate points I wanted to make, and motivated the writing. I hope you enjoyed the result.

Index

A
Alexanderson alternator, 121
Alexanderson, Ernst, 1, 27, 29, 99
American Marconi, 10, 27, 48, 121–122
American Telephone & Telegraph Company (AT&T), 27, 48, 49, 64, 122
Anderson, John, 123–124, 129, 134
Anglo-American Telegraph Company, 8
Armstrong, Edwin, H., 52, 56, 67, 69, 71, 76, 98, 103–104

B
Berliner, Emile, 58
Bell Telephone Company, 31, 47–50, 60–61, 64–66
Bell, Alexander Graham, 8, 48–49
Bell, Melville, 48
Belle Isle, 12, 34–35, 43–44

C
Californian, 7
Canadian De Forest patents, 85
Canadian General Electric Company (CGE), 29, 31, 51, 58, 63, 71, 87, 88, 117, 121, 125–126, 131–134, 136, 139–140
Canadian Government, 7, 9–10, 12–15, 17, 20, 23, 28, 30–32, 39, 43, 97, 105
Canadian Independent Telephone Co. Limited (CITCo), 48, 65–67, 69–72, 75–76
Canadian Intellectual Property Office, 98
Canadian Marconi Company (CMC), 7, 11, 24, 28, 33–34, 40–41, 45, 51, 87, 117, 121
Canadian McCullough AC Radio Tubes, 80–81
Canadian National Railway (CNR), 36, 38–40
Canadian Pacific Railway (CPR), 36, 46
Canadian Radio Corporation, 71, 75, 87
Canadian Radio License Agreement, 31, 36
Canadian Radio Patents Ltd., 37, 87, 113, 117
Cape Cod, Massachussets, 8, 30
Cape Race, 12, 18, 21, 31, 44
carborundum detector, 5–6, 21
Cartmel, W.B., 55–56, 61, 63
Cdn Westinghouse, 31, 37, 50, 63, 87, 103–104, 106, 107, 117, 121, 124–134
Cdn Westinghouse Aeriola RS, 125
Cdn Westinghouse Aeriola Sr, 125, 127
Cdn Westinghouse Battery Operated, 125, 130, 131
Cdn Westinghouse Batteryless Receiver, 125, 131, 132
Cdn Westinghouse Model 8, 125, 129
Cdn Westinghouse Model 16, 125, 130, 131
Cdn Westinghouse Model 53, 126, 133
Cdn Westinghouse Model 55, 126, 133, 134
Cdn Westinghouse Model 56, 126
Cdn Westinghouse Model 57 Special, 126, 133
Cdn Westinghouse Model 58, 126
Cdn Westinghouse Model 60, 126, 133, 134
Cdn Westinghouse Model 69, 126
Cdn Westinghouse Model 89, 126
Cdn Westinghouse Model 99, 126
Cdn Westinghouse Radiola AC, 125
Cdn Westinghouse Radiola Bal. Amp, 125, 128
Cdn Westinghouse Radiola III, 125, 127–129, 136, 139
Cdn Westinghouse Radiola IIIA, 125, 129
Cdn Westinghouse Radiola Regenoflex, 125, 132
Cdn Westinghouse Radiola Super, 125, 129, 142
Cdn Westinghouse Radiola X, 125, 129, 130
Cdn Westinghouse RC, 125
CFCA, 71, 72
CFCF, 31, 34–40
CFCN, 104–105, 107–108
CFRB, 88, 91
CGE Model 93, 193, 126, 134, 139
CGE Model 118, 125, 127
CGE Model 205, 126, 134, 139
CGE Radiola 16, 125, 130, 132
CGE Radiola 17, 125, 131, 132
CGE Radiola 18, 125, 131, 132
CGE Radiola 20, 125, 132
CGE Radiola 21, 125
CGE Radiola 22, 125
CGE Radiola 24, 125, 139
CGE Radiola 25, 125, 139
CGE Radiola 28, 126, 129, 139
CGE Radiola 30-A, 126
CGE Radiola 32, 126
CGE Radiola 33, 126
CGE Radiola 44, 126
CGE Radiola 46, 126
CGE Radiola 47, 126
CGE Radiola 50, 126, 132
CGE Radiola 51, 126, 132, 133
CGE Radiola 60, 126
CGE Radiola 62, 126
CGE Radiola 64, 126
CGE Radiola 66, 126
CGE Radiola 67, 126
CGE Radiola III, 125, 128, 139
CGE Radiola IIIA, 125, 129
CGE Radiola Super VIII, 125
CGE Radiola Super. AR812, 125, 129, 130, 139
CHYC, 58
CITCo Everyman crystal set, 66, 110
CITCo SC-10 receiver, 67
CITCo Type S amplifier, 67
CITCo Universal Receiving Set, 66–67
CKAC, 31, 58
CKCE, 65, 71
clapper brake, 104
Clifden, 13, 18, 44
CMC Arcon Junior, 25–26
CMC Model 11, 33, 38
CMC Model 12, 33, 38
CMC Model 13, 33, 38
CMC Model 14, 33, 39–40

CMC Model 15, 33, 40
CMC Model 16, 33, 40
CMC Model 17, 33, 40
CMC Model 19, 33, 40
CMC Model 20, 33, 40
CMC Model 21, 33, 40
CMC Model 22, 33, 40
CMC Model 23, 33, 40
CMC Model 24, 33, 40
CMC Model 26, 33, 39–40
CMC Model 28, 33, 40
CMC Q valve, 26
CMC V24 valves, 25–26
Coats, D.R.P. (Darby), 7, 22, 24, 29
coherers, 43–44
continuous wave, 1–2, 6, 30–31, 44–45
Cowherd, James, 48
cyclic heating factor, 73

D

damped wave, 45
De Forest's audion, 48, 100–102, 110
De Forest Radio Corporation of Canada, 118
De Forest Radio Telephone and Telegraph Company, 100–101
De Forest, Lee, 8, 25, 44, 48, 65–67, 69, 71, 76, 82, 91, 98–100, 102, 110
Diamond Jubilee, 36

E

E.I. Co., 3–4, 6, 110, 112
Eaton, T., Co., 76, 109–112, 139, 143–144
Eaton Single Circuit Set, 109–110
electrolytic detector, 3–6
Evans, Cyril "Corp", 7
Evans, Mathew, 98

F

Fessenden, Reginald, 1–3, 5–6, 8, 25, 43–45, 98–99
Fielding, W.S., 9
Fleming, John Ambrose, 44, 99–100
Fraser, A.F., 7, 43
Freeman, Hubert W., 74, 91

G

galena detector, 5, 6
General Electric (GE), 27, 44, 48, 91, 121, 123, 127, 132, 134, 139
Gillis, Gar, 143, 146
Given, T.H., 1
Glace Bay, 8–9, 11–15, 17–18, 23, 30, 31, 36, 39, 43–45
Grant Electronics Ltd, 146
Grant Perfection, 105–106
Grant Superdyne, 105–106
Grant, W.W., crystal set, 105, 107
Grant, W.W., Radio Ltd., 103, 105–107, 146
Grant, W.W., Voice of the Prairie, 103, 105–107
Grant, W.W., Voice of the Prairie Five, 106
Grant, W.W., Voice of the Prairie Four, 107
Grant, W.W., Voice of the Prairie Seven, 107
Grant, William Walter, 103–108, 117, 146

H

Hammond Manufacturing Co, 97
Hammond Museum, 54, 64, 94
Hammond, Fred, 64, 94, 97
heterodyne reception, 44–45, 99, 102
Hull, Dr. A.W., 74, 80

I

Inspection Board of U.K. & Canada, 141
International Western Electric Co., 31, 50
IXCR, 137–138

K

KDKA, 24–25
Kellogg, 78
Kellogg 401, 82
Kemp, George, 8
Kipp, H.M., 113, 116, 118
Kipp, H.M. Company, 113, 116–117, 140

L

Laurier, Sir Wilfred, 9
Levy, Lucien, 56
liquid barretter, 1–3, 6
Love, H. Gordon, 103–104
Lowry, Charles A., 113, 116–119, 140

M

magnetic detector, 21, 44, 99
magnetic field, 73, 74
Marconi "C" set, 26
Marconi Company, 7, 10–11, 13–14, 17, 19–21, 23–28, 31, 33–34, 41, 45, 47–48, 51, 58, 74, 87, 101–102, 117, 121–122, 137
Marconi International Marine Communications Company Limited (MIMCo), 7, 9–10, 17, 20
Marconi, Guglielmo, 7–10, 13–14, 27, 33, 43–44, 69, 99–100, 102, 137
Marconi's Wireless Telegraph Company Limited (MWT), 9–15, 23–24, 27, 29, 31, 34, 36–37, 50
Marconiphone Mk I, 32, 33
Marconiphone Mk II, 32, 33
Marconiphone Mk III, 32, 33
Marconiphone Mk IV, 33
Marconiphone Mk V, 32–33
Marconiphone Mk VI, 33, 35
Marconiphone Mk VII, 33, 35
Marconiphone Mk VIII, 33, 36
Marconiphone Mk IX, 33, 36
Marconiphone Mk X, 33, 37
marine communications, 10, 29
Marine Department, CMC, 29, 32, 37, 39, 44
McCullough 401 tube, 79, 82
McCullough Sales Company, 76, 78
McCullough Tubes, 67, 69, 76–77, 79–81, 117
McCullough, Frederick, 66, 75–79, 81, 85, 91
Mercury Jr., 117–118
Mercury Super Ten, 113–114, 116

N

Nally, Edward J., 122–123
National Electric Signaling Company, 1
NE DX-235, 61

NE R-1,   51–52, 56
NE R-2 Short Wave Tuner,   51, 53–56
NE R-3,   51
NE R-3 Four-tube Radio-frequency Set,   54, 60
NE R-4 Superheterodyne Radio Receiving Set,   51, 54–56, 59–61
NE R-11 Radio Receiving Set,   51, 55–56
NE R-4-L,   51
NE R-5/R-5A,   51, 52
NE R-12 Radio Receiving Set,   51, 56, 59
NE R-15 Amplifier,   51, 55–56
NE R-20,   51, 52
NE R-21,   51
NE R-22,   51
NE R-23,   51
NE R-24,   51
NE R-30,   51
NE R-31,   51
NE R-40,   51
NE R-41,   51
NE R-50,   51
NE R-105,   51, 52
NE R-106 Loud Speaker Amplifier,   51, 57–58
NE R-107 Loudspeaker Amplifier,   51, 58
NE R-203-D,   61
NE R-208-A,   61
NE R-215-A,   61
NE R-216-A,   61
NE R221-D,   61
NE R-221-DX,   61
NE R500 speaker,   63
NE R518 Loudspeaking Receiver,   63
NE R540 cone speaker,   63
NE R-1000 Radio Receiving Set,   52
NE R6900 Loud Speaker,   63
Northern Electric and Manufacturing Company, Limited,   49
Northern Electric Company Ltd (NE),   31, 37, 47–52, 55–56, 58, 64, 87, 105, 117, 140

P
Paget, Percy   8
perikon detector,   5–6, 124
platinum wire,   2–4
Poldhu, England,   7–9
Port Morien,   12

Q
QRS Music Company of Canada Ltd.,   81–82, 84, 92

R
Radio College of Canada (RCC),   94, 117–118, 124–126, 128–129, 132, 134, 144, 146
Radio Corporation of America (RCA),   27–28, 36, 38, 58, 60, 69, 89, 91, 101, 104, 113, 121–135, 139–140, 143
Radio Devices Limited,   116–117, 119
Radio Devices Limited. Model K crystal set,   116–119
Radio Music Box,   48
Radio Valve Company of Canada Ltd,   31
Receiver Manufacturing and Design Committee,   123
Richardson, Charles Gordon,   136–137
Richardson, Charles Lepel,   121, 136–140, 142
Riders manuals,   124–126, 128, 132

Rogers "Jubilee" chassis,   89, 91–92
Rogers 100- 110- 120 Circuit,   85
Rogers AC-20,   86, 88, 91
Rogers AC-30,   88–89
Rogers AC-32,   82, 85–86, 88, 90–91
Rogers Batteryless,   81, 86–88, 125, 131–132
Rogers Majestic,   31
Rogers Model 20,   82
Rogers Model 30,   82
Rogers Model 50,   84, 86
Rogers Model 90,   88, 90
Rogers Model 120,   85–86, 88
Rogers Model 130,   85–86
Rogers Model 135,   86
Rogers Model 200,   88
Rogers Model 220,   88
Rogers Model 250,   88–91
Rogers Model 420,   93
Rogers Model 480,   93–94
Rogers Patents,   80
Rogers R-15,   89
Rogers R-100,   79, 84, 86
Rogers R-200,   79, 86–87
Rogers Radio Company,   67, 69, 75–76, 79–94, 117, 143
Rogers RX-100,   79
Rogers RX-200,   79
Rogers, Albert. S.,   75–76, 91
Rogers, David,   82
Rogers, Edward S. (Ted),   48, 66–67, 69–70, 72, 75–77, 79–82, 85–91, 131
Rogers, F.S.,   76
Rogers, Samuel,   76
Rogers/Standard Radio,   87
Rogers-Majestic Corp. Ltd.,   91, 140
Round, H.J.   74

S
Salton, Lynn V.,   109–110
Sarnoff, David,   38, 48, 70, 122
Scientific Experimenter Limited,   25–27
short-wave "beam" stations,   34–36, 38–39, 45–46
Sise, Charles F.,   49, 58
SkyWaves crystal set,   3–5
spark equipment,   30, 45
St. John's, Newfoundland,   7–8, 22, 43–44
Standard Radio Manufacturing Corp. Ltd.,   37, 80–81–82, 84–85, 87, 89, 91
Stokes, John,   61
superheterodyne,   35, 54–55, 59–60, 106–107, 113, 129, 139

T
Table Head,   8, 10, 12
Thompson, Roy,   47
Thorcraft,   143–147
Thorkelsson, Hank,   143, 147
Titanic, S.S.   7, 18, 20, 70
Tyne, Gerald,   61, 69

U
U.S. Navy Department,   4, 6, 27, 121–122
United Fruit Company,   1, 6, 122

V

Victor Talking Machine Company of Canada Limited, 51, 58–60, 63
Victor/Northern Electric, 59–60, 63
Victor/Northern Electric type R-20, 51–52, 59
Victor/Northern Electric type R-50, 60–61
voltage gradient, 73, 74
Vyvyan, Richard 13

W

Wade, Wallace, 74, 91
Walker, Hay, 1
Watson, Charles, 48
Western Electric Company, 27, 31, 48–51, 54–55, 58, 60–64, 74, 122–123
Westinghouse Company, 1, 5, 27, 31, 66, 69, 74–75, 77–78, 91, 121, 122–135, 139
Westinghouse, George, 1
Whitehead and Richardson, 139
Williams Jr., Charles, 48, 49
Williams, Ralph, 52
Wireless Specialty Apparatus Company (WSA) 122–123, 125
Wolloston wire, 2
Woodward, Dr. Henry, 98

X

XWA, 7, 23–25, 28–29

Y

Young, Owen, 38, 122

Marconi's History

GIAN CARLO CORAZZA, LIFE ASSOCIATE MEMBER, IEEE

The roots of the future are in the past.

This paper remembers the main events of the history of radio, events in which Guglielmo Marconi gave his fundamental contribution to the development of the new system of communication.

This paper covers the period from the first transmission at Villa Griffone, near Bologna (Italy), in 1895, to the transmission across the Atlantic Ocean from Poldhu (Cornwall) to St. John's (Newfoundland) in 1901.

Keywords—History, radio communication.

Guglielmo Marconi was born in Bologna, Italy, on April 25, 1874. Therefore, he was only 21 years old when, in the spring of 1895, at Villa Griffone in Pontecchio (Bologna), he transmitted the first signal using a free propagating electromagnetic wave as a carrier.

However, radio was not born in one day. Rather, it had to go through a long labor, which ended in 1901, at Signal Hill (St. John's, Newfoundland), with the first transmission across the Atlantic Ocean. We can reasonably hypothesize that if the activity Marconi carried out between 1896 and 1901 had not been successful, the 1895 experiments would have only opened the way to plain radio telegraphy. In fact, as we will further discuss below, in 1895, the instrumentation was not syntonic, and the ionosphere had not been discovered yet.

As a consequence, the words "invention of radio" do not identify a single event but rather a whole period.

The scenery where the initial part of the history of radio communication took place is today known as *Villa Griffone* (Fig. 1), but in Marconi's times it was called *Il Griffone*. The building, elegant and impressive, was actually a large country mansion with a residential part, a stable, a barn, and a granary where Marconi's father reluctantly let him build the first laboratory, thanks to the insistent requests of his mother, Annie Jameson.

During the autumn–winter 1894–1895, Marconi worked in his laboratory with the aim of putting into practice what he had learned from the activity of other researchers.

In the spring of 1895, Marconi opened the windows of the granary and brought the receiver outside the building, further and further away from the transmitter that was left inside.

Manuscript received June 18, 1997; revised March 4, 1998.
The author is with the University of Bologna, Bologna 9-40135 Italy.
Publisher Item Identifier S 0018-9219(98)04467-3.

Fig. 1. Villa Griffone, Pontecchio Marconi, Italy.

Once he had placed also the transmitter outside Villa Griffone, Marconi resumed his experiments until he had the stroke of genius of a great inventor. At the conference of Stockholm, when in 1909 he received the Nobel Prize, he said:

> In August 1895 I hit upon a new arrangement which not only greatly increased the distance over which I could communicate but also seemed to make the transmission independent from the effects of intervening obstacles.
>
> This arrangement [Figs. 2 and 7] consisted in connecting one terminal of the Hertzian oscillator, or spark producer to earth and the other terminal to a wire or capacity aerea placed at a height above the ground and in also connecting at the receiving end [Figs. 3 and 6] one terminal of the coherer to earth and the other to an elevated conductor. [1]

The new system immediately proved extremely effective, either to increase the distance between the transmitter and the receiver or "to overcome hills, mountains, large metallic obstacles ... intervening between the places where the link must be established."

Thanks to the new arrangement, the distance between the transmitter and receiver was progressively increased, to the point that the inventor's assistants finally found themselves at the upslope of the *Celestini Hill* (Fig. 4), situated about one mile from Villa Griffone. According to

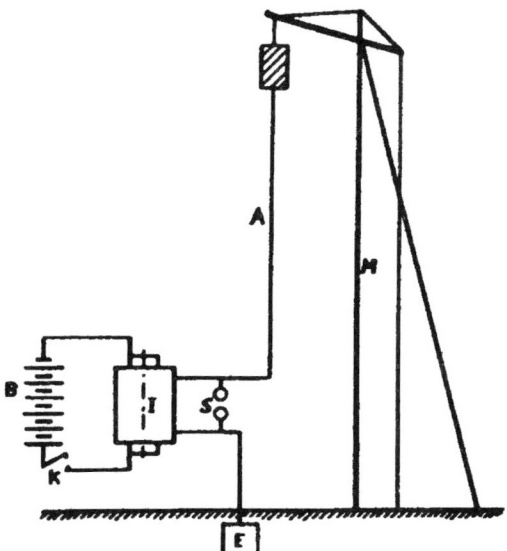

Fig. 2. Scheme of the transmitter used by Marconi at Villa Griffone (1895).

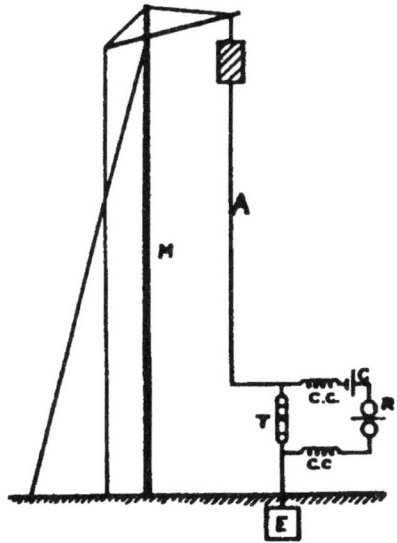

Fig. 3. Scheme of the receiver used by Marconi at Villa Griffone (1895).

what was reported in an interview published in 1897 by English journalist H. J. W. Dam, Marconi declared that in September 1895, one of his receivers, which had been placed on the other side of the hill, reacted to the signals transmitted from Villa Griffone.

During 1896, Marconi moved to Great Britain, where on June 2, he filed a patent entitled "Improvements in Transmitting Electrical Impulses and Signals, and in Apparatus Thereof."

The first practical demonstrations took place in London, in summer 1896, and in September of the same year in the plain of Salisbury. The technical report that H. R. Kempe of the Engineer in Chiefs Office sent to the General Post Office on the Marconi system reads:

> The experiments which were made with the above system on Salisbury Downs and which were concluded on Friday last have to a very great extent been successful ones and I am certain by now that the system is likely to prove of considerable value when further developed. In many respects the results obtained were most remarkable. [2]

The tests of Salisbury were followed by a long series of experiments and applications, which testify to the huge amount of work that was necessary to achieve the complete invention.

In May 1897, some messages were exchanged between *Lavernock Point* and the *Isle of Flatholm*, in the Bristol Channel, at a distance of nine miles.

In the same year, Marconi came back to Italy and carried out some experiments in Rome and at La Spezia. In the Gulf of La Spezia in particular, he obtained satisfactory results up to a distance of 18 km on sea, between *Cape s. Bartolomeo* and the battleship *San Martino*.

Assisted by the Wireless Telegraph and Signal Co., Ltd., a company he had founded with a group of English financiers in July 1897, Marconi began to build two telegraphic stations, one at *Alum Bay*, in the Isle of Wight, and the other, 14 miles distant, at *Bournemouth*, in the village of *Poole*. The same distance was also covered while transmitting from the Isle of Wight to a steamboat. This last experiment was evidence that bad weather conditions do not prevent successful communications.

During May 1898, upon a request by the Lloyds, radio transmissions took place between *Ballycastle* and the isle of *Rathlin*.

On July 1898, aboard the boat *Flying Huntress*, Marconi followed the Kingstown regatta for the *Daily Express* of Dublin, and he sent messages that were afterwards published in the evening edition of the newspaper. A connection was also established between the boat *Osborne* and *Osborne House*, in the Isle of White, to permit Queen Victoria to communicate with her son, the Prince of Wales. In the last case, about 1000 messages were transmitted in a satisfactory way, and Marconi commented:

> I consider the results of this service particularly interesting, because many people doubted the possibility of establishing regular radiotelegraphic communications over long stripes of land. [3]

The ease of use and the reliability of the apparatus were again underlined in the connection between the lightship *East Goodwin* (Fig. 5)—moored 12 miles off the lighthouse of *South Foreland*—and the lighthouse itself. On March 3, 1899, the *East Goodwin* accidentally collided with a steamboat that was sailing in fog, and the S.O.S. message received at South Foreland allowed the rescue of the entire crew.

A regular radio-telegraphic service between England and France was opened on March 27, 1899, between the station installed in the lighthouse of *South Foreland* and the station of *Chalet l'Artois*, at *Wimereux*, not far from Boulogne, where Marconi's parents had married.

Fig. 4. View from the rear of Villa Griffone, toward the west. Celestini Hill is on the left.

Fig. 5. The lightship *East Goodwin*.

Fig. 6. The first receiver.

Other tests were carried out in 1899: a service for two American journals, on the occasion of the America's Cup regatta; and a transmission between two ships of the U.S. fleet—the *New York* and the *Massachusetts*.

Finally, in 1900, Marconi managed to accomplish multiple transmission on a single apparatus by exploiting the phenomenon of syntony or tuning and he had, once again, a stroke of genius.

The text of the conference Marconi held at the Society of Arts on May 15, 1901, reads:

> I now wish to describe ... further improvements I carried out, regarding in particular the results I obtained while tuning or syntonizing the apparatus.
> As long as it was possible to operate only two apparatus within what I'll call the sphere of their influence, a great limit was set to the practical use of the system.
>
> Using simple wires, placed vertically, connected directly to the coherer and to the spark gap of receiver and transmitter, as I used to do before 1898, it was not possible to get any satisfactory syntony. The new tuning methods which I adopted in 1898 [Fig. 6] consisted in the connection of the receiving aerial to the ground instead of connecting it to the coherer, and in introducing a suitable form of oscillations transformer coupled with a condenser; in this way I had built a tuned resonator which better answered to the waves emitted by an aerial of a determined length. [4]

Soon, also the transmitter was changed (Fig. 7), until its configuration became very similar to the receiver previously described, and the whole system—transmitter and receiver—became syntonic.

At the Nobel Prize conference in Stockholm, Marconi affirmed:

Fig. 7. The first transmitter.

Fig. 8. View of Signal Hill from the sea.

In 1900 I constructed and patented (it is the patent N. 7777 issued on 26 April) transmitters which consisted of the usual kind of elevated capacity aerea and earth connection, but this was inductively coupled to an oscillation circuit containing a condenser, an inductance and a spark gap, the conditions which I found essential for efficiency being that the periods of electrical oscillation of the elevated wire or conductor should be in tune or resonance with that of the condenser circuit.

At the end of 1900, Marconi was ready to look at the most ambitious goal, and the reading of the conference of Stockholm becomes fascinating. Here are his words.

In January 1901 some successful experiments were carried out between two points on the South Coast of England 186 miles apart, i.e., St. Caterine's Point (Isle of Wight), and The Lizard in Cornwall.

The total height of these stations above the sea level did not exceed 100 meters, whereas to clear the curvature of the earth a height of more than 1600 meters at each end would have been necessary.

The results obtained from these tests ... seemed to indicate that electric waves produced in the manner I had adopted, would most probably be able to make their way round the curvature of the earth, and that therefore even at great distances, such as those dividing America from Europe, the factor of the earth's curvature would not constitute an insurmountable barrier to the extension of Telegraphy through space.

The belief that the curvature of the earth would not stop the propagation of the waves, and the success obtained by syntonic methods in preventing mutual interference led me in 1900 to decide to attempt the experiment of testing whether or not it would be possible to detect electric waves over a distance of 4000 kilometers, which, if successful, would have immediately proved the possibility of telegraphing without wires between Europe and America.

Today we know about the existence of the ionosphere and the fundamental role it had in the Transatlantic transmission, but in 1901 it was completely unknown. All the analyses carried out by electromagnetic field experts of the time, based on an incomplete propagation model, resulted in the impossibility of the closure of such an extremely long link. If Marconi had not been a strong-minded man and an ingenious experimenter, he would have probably quit the challenge and would not have discovered the physical phenomenon of the ionosphere. But how much struggle and hard work to reach the final results.

First, Marconi and his staff built up the transmitter station in Poldhu (Cornwall), with about 15 kW of power for a nominal wavelength of 1800 m. It must be noted, however, that this transmitter had a large spurious harmonic content; therefore, Marconi was certainly using also wavelengths shorter than 1800 m.

The receiver was in St. John's, Newfoundland, at a distance from Poldhu of 3684 km on top of a hill baptized with the prophetic name *Signal Hill* (Fig. 8) because it was used for flag signaling to the ships. During the historic days of the experiment, a blustering wind prevented the use of balloons to hold the receiving antenna wire vertical. At the end, Marconi decided to use a kite (Fig. 9) that soared, tied to a copper wire 120 m long, which was actually the antenna. The receiving apparatus was again nonsyntonic and used as a detector a "mercury-drop" or "Italian Navy coherer" and earphones in order to increase sensitivity. On December 12, 1901, around 30 minutes past noon, three weak but detectable signals transmitted from Poldhu reached Marconi's ears (Fig. 10). He handed the earphones to his trusted assistant Kemp to have a confirmation of what he had heard, asking him: "Can you hear anything,

Fig. 9. Marconi (left) and his assistants at Signal Hill (1901).

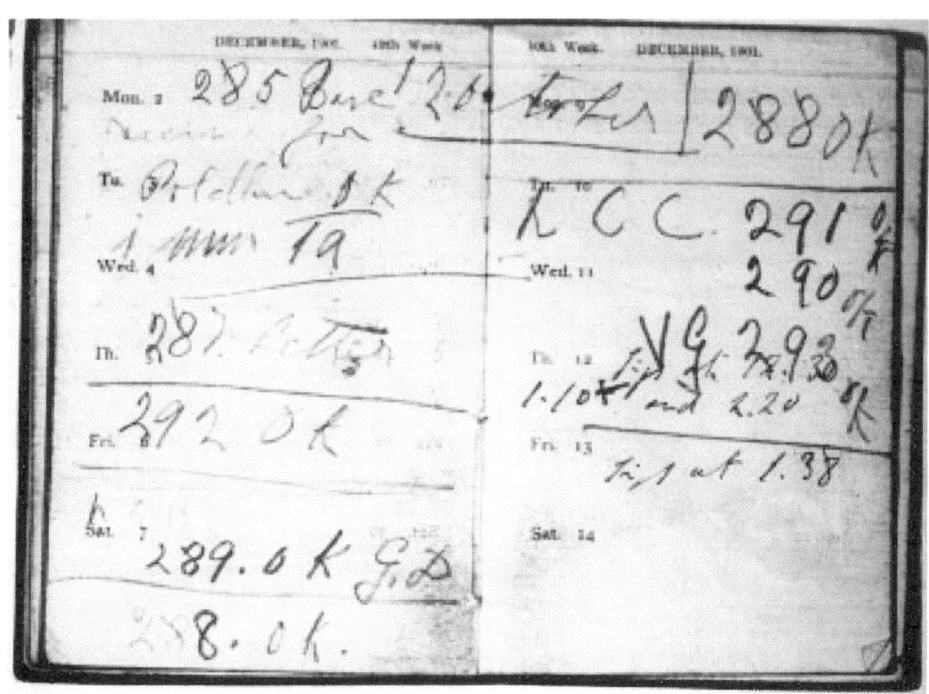

Fig. 10. The agenda where Marconi wrote he received a signal at 12:30, 1:10, and 2:20 on December 12, 1901.

Mr. Kemp?" The positive answer to this simple question started the age of Marconi's heritage, the age of radio, the age of wireless communications.

REFERENCES

[1] G. Marconi, "Original Manuskriptet Marconis Nobelpris-Föredrag," Hallet Inför Kungl. Svenska Vetenskapsakademien, den Dec. 11, 1909.
[2] H. R. Kempe, "Cooperation of department with Mr. Marconi, experiments on Salisbury Downs," General Post Office, London, UK, September 1896.
[3] G. Marconi, presented at the Institution of Electrical Engineering Conf., London, UK, Mar. 2, 1899.
[4] ——, presented at the Society of Arts Conf., London, UK, May 15, 1907.

Gian Carlo Corazza (Life Associate Member, IEEE) received the industrial engineering degree from the University of Bologna, Bologna, Italy, in 1951.

He joined the Istituto Superiore P.T., Rome, Italy, in 1953. He currently is Professor of electromagnetic fields at the University of Bologna. He is a past President of the Fondazione Guglielmo Marconi of Pontecchio Marconi, Italy.